INTRODUCTION TO
COMPLEX
VARIABLES

E. A. GROVE

G. LADAS

Department of Mathematics
University of Rhode Island

Houghton Mifflin Company   Boston

Atlanta   Dallas   Geneva, Illinois   Hopewell, New Jersey   Palo Alto   London

Printed in the U.S.A.

Library of Congress Catalog Card Number: 73-9413

ISBN: 0-395-17087-7

CONTENTS

PREFACE

The theory of functions of a complex variable plays a central role in both pure and applied mathematical analysis. It is a remarkably harmonious theory, one that has great applications in such diverse fields as the theory of equations, conformal mappings, hydrodynamics, aerodynamics, thermodynamics, potential theory, non-Euclidean geometry, and topology.

An elementary course in complex variables will enable us to solve algebraic equations and to give a simple proof of the fundamental theorem of algebra (which states that a polynomial of degree $n \geq 1$ has at least one complex root). We shall get a better understanding of the concept of convergent power series. The residue theory will enable us to compute some quite useful real integrals (whose computation is extremely difficult without the theory of complex variables). After the basic theory of complex variables has been developed (something of great mathematical interest in itself), several applications to problems of mathematical physics, differential equations, and engineering will be given.

The first five chapters of this book represent what the authors consider must be mastered by any undergraduate mathematician, physicist, or scientific engineer today. Chapter 6 provides a brief glimpse of some of the various applications of complex variables to other branches of mathematics. Finally, the appendices contain a resumé of some material of a more advanced nature (as well as of some that is not) which should whet the reader's appetite for further study of the subject.

This book is intended to be used for a one-semester course in complex variables offered to undergraduates or first-year graduate students in mathematics, physics, and engineering.

The prerequisite for this book is undergraduate calculus. All other needed results have been developed.

While class-testing this material, we have frequently found it convenient to omit the proofs of some theorems in Chapter 2 and the proofs of the generalized Cauchy theorems in order to concentrate more heavily on the applications of the Cauchy residue theory.

We would like to thank our teachers Professors H. Federer, P. Garabedian, and D. Kappos who taught us the subject. Thanks are also due to the following people who read the manuscript with great care and made many helpful suggestions concerning content, presentation of certain proofs, and overall clarity: Klaus Bichteler, University of Texas at Austin; Herman Flaschka, University of Arizona; Simon Hellerstein, University of Wisconsin; Thomas McCoy, Michigan State University; Alexander Poularikas, University of Rhode Island; John Scheik, Ohio State University; Rod Smart, University of Wisconsin; and Lawrence Zaleman, Stanford University. Special thanks must go to A. Poularikas for contributing Secs. 6.4 and

6.5 of the applications. We wish to thank our student Linda Helnes for the excellent typing of the manuscript, the staff of Houghton Mifflin for the expert advice and encouragement they gave, and finally our wives and children for their patience during the long evening hours we spent working on the manuscript.

E. A. Grove

G. Ladas

INTRODUCTION TO
COMPLEX
VARIABLES

Historical Remarks

The imaginary numbers were invented during the 16th century when mathematicians sought a general solution of quadratic and cubic equations. Since the square of every real number is either positive or zero, the equation $x^2 = -1$ cannot be solved in the field of real numbers. At the beginning, complex numbers were developed by adjoining the symbol $\sqrt{-1}$ to the real number system. This symbol, however, is not satisfactory and leads to paradoxes such as the following:

$$-1 = (\sqrt{-1})^2 = \sqrt{-1}\,\sqrt{-1} = \sqrt{(-1)(-1)} = \sqrt{1} = 1$$

To avoid paradoxes like this, Leonhard Euler (1707–1783, Swiss) introduced in 1777 the notation i with the basic property $i^2 = -1$. The two roots of the equation $x^2 = -1$ are now $\pm i$. The symbol i is called the *imaginary identity*. The choice of the word *imaginary* is unfortunate, but it indicates the distrust with which complex numbers were viewed. These suspicions slowly vanished at the end of the 18th century, when Caspar Wessel (1745–1818, Norwegian) in 1797 and Carl Friedrich Gauss (1777–1855, German) in his doctoral thesis in 1799 gave a simple geometric representation to complex numbers $a + ib$. Wessel and Gauss thought of a and b as representing the rectangular coordinates of a point in the Cartesian plane $\mathbf{R} \times \mathbf{R}$ ($\mathbf{R}$ standing for the real line). This simple interpretation of complex numbers made mathematicians feel much more comfortable about imaginary numbers, and their existence was slowly accepted. In 1833, Sir William Rowan Hamilton (1805–1865, Irish) presented a paper before the Royal Irish Academy in which he introduced a formal algebra of ordered pairs of real numbers, the rules of combination being precisely those given today for the system of complex numbers.

In the 18th century, Euler began the study of functions and series of a complex variable. He observed that the formal substitution of x by ix in the exponential function

$$e^x = 1 + \frac{x}{1!} + \frac{x^2}{2!} + \frac{x^3}{3!} + \cdots$$

leads to

$$e^{ix} = \left(1 - \frac{x^2}{2!} + \frac{x^4}{4!} - \cdots\right) + i\left(x - \frac{x^3}{3!} + \frac{x^5}{5!} - \cdots\right)$$

that is,

$$e^{ix} = \cos x + i \sin x$$

Similar methods led to other striking results. However, all these "formal" results were lacking in mathematical rigor and often led to paradoxes. For example, we know that the real-valued function $y = \tan x$ for $-\pi/2 < x < \pi/2$ takes on all real values. Suppose that this function could be generalized so that it could take on all complex values while the ordinary law of the tangent of sums remained valid. There should then exist a complex number x_0 such that $\tan x_0 = -i$. Thus for any complex number x with $\tan x \neq \pm i$, we should have

$$\tan (x + x_0) = \frac{\tan x + \tan x_0}{1 - \tan x \tan x_0} = \frac{\tan x - i}{1 + i \tan x} = -i$$

which is absurd.

It was not until the 19th century that this naïve approach to complex analysis was replaced by a rigorous treatment. The founders of the theory of functions of one complex variable, and of all analysis, were Augustin Louis Cauchy (1789–1857, French), professor at the École Polytechnique in Paris in 1848, Karl Weierstrass (1815–1897, German), professor at the University of Berlin in 1864, and Bernhard Riemann (1826–1866, German), professor in Göttingen in 1859. Cauchy introduced the concept of the complex line integral in 1814 and published his basic theorems on functions of one complex variable in 1825. During the second half of the 19th century, Riemann developed the theory of complex functions from a physico-geometrical standpoint, and Weierstrass developed it from a logically rigorous standpoint.

The invention of set theory by Georg Cantor (1845–1918, St. Petersburg) at the end of the 19th century helped enormously in the development of the foundations of complex analysis, and it was widely used.

Among the numerous mathematicians who contributed to the application of complex analysis to other branches of mathematics, such as potential theory (Dirichlet problem) and conformal mappings, we should mention Hermann Amandus Schwarz (1843–1921, German), a student and successor of Weierstrass.

Chapter 1   The Algebra of
Complex Numbers

1.1
INTRODUCTION

We begin by defining complex numbers in the same way that they were defined by Hamilton in 1833 (see the biography of Hamilton), namely as ordered pairs (a,b) of real numbers obeying certain algebraic operations. This definition has the advantage of showing the beginner that there is nothing unreal about the so-called "imaginary" numbers. Hamilton's approach is also used today when complex numbers are utilized in computer programming. Next we establish that Hamilton's notation (a,b) and the older notation of Euler $a + ib$, where $i^2 = -1$, are equivalent. Finally we develop the algebra of complex numbers and their geometric representation (as understood since 1799 by Gauss); we also present some simple applications of complex numbers to solutions of algebraic equations.

1.2
COMPLEX NUMBERS

We start with an axiomatic definition of the complex numbers. The complex numbers are the set **C** of all ordered pairs (a,b) of real numbers together with the two operations of addition, $+$, and multiplication, $\cdot$, defined as follows:

$$(a,b) + (c,d) = (a + c, b + d)$$

$$(a,b) \cdot (c,d) = (ac - bd, ad + bc)$$

Two complex numbers (a,b) and (c,d) are equal if and only if $a = c$ and $b = d$.

It is easily seen that the set $\mathbf{C}$ of complex numbers is a *field* with respect to the above two operations of addition and multiplication, with the *zero of*

WILLIAM ROWAN HAMILTON

William Rowan Hamilton was born in 1805 in Dublin, Ireland, where his father was a practicing attorney. When William was 12, his mother died, and two years later his father also died. However, his education had been undertaken long before by his uncle James Hamilton, who was a linguist. Due to his uncle's efforts, William mastered 13 languages by the time he was 13 years old. More important to his future career, his mathematical interest was awakened by a reading of Newton's *Arithmetica Universalis* when he was only 12. Soon after, he read the *Principia*. The American lightning-calculating boy, Zerah Colburn, also greatly influenced Hamilton's interest in mathematics.

Hamilton was still an undergraduate at Trinity College, Dublin, when he was appointed professor of astronomy of that prestigious college. At the age of 28, he presented to the Royal Irish Academy a significant paper in which he introduced a formal algebra of pairs (a,b) of real numbers which is actually the definition of complex numbers that we present in Sec. 1.2.

Two unhappy love affairs (he attempted to drown himself after the first one), a hypochondriac wife, and alchoholism marred the personal life of this great Irish mathematician. Although it is thought by many that these difficulties lowered the quality of his mathematical thought, Hamilton's mathematical output continued unabated to the end of his life, with a number of published papers. His *Lectures on Quaternions* was published in 1853, and he was working on his great *Elements of Quaternions* at the time of his death in 1865. The book was published posthumously but did not have the influence on mathematics that Hamilton had hoped. An enormous collection of other manuscripts and notes was turned over to Trinity College.

---

addition being the complex number $(0,0)$, the *identity of multiplication* being the complex number $(1,0)$, the *negative* of (a,b) being $(-a,-b)$, and the *inverse* of $(a,b) \neq (0,0)$ being

$$(a,b)^{-1} = \left(\frac{a}{a^2 + b^2}, \frac{-b}{a^2 + b^2} \right)$$

To prove the above statement, let $z_1 = (a,b)$, $z_2 = (c,d)$, and $z_3 = (e,f)$ be arbitrary complex numbers. All that we have to show is

1. $z_1 + z_2 = z_2 + z_1$ and $z_1 \cdot z_2 = z_2 \cdot z_1$. That is, addition and multiplication are *commutative*.
2. $z_1 + (z_2 + z_3) = (z_1 + z_2) + z_3$ and $z_1 \cdot (z_2 \cdot z_3) = (z_1 \cdot z_2) \cdot z_3$. Addition and multiplication are *associative*.
3. $z_1 + (0,0) = z_1$ and $z_1 \cdot (1,0) = z_1$.
4. $z_1 + (-z_1) = (0,0)$ and $z_2 \cdot z_2^{-1} = (1,0)$ if $z_2 \neq (0,0)$. Here $-z_1$ denotes the negative of z_1 and z_2^{-1} the inverse of z_2.
5. $z_1 \cdot (z_2 + z_3) = z_1 \cdot z_2 + z_1 \cdot z_3$. That is, multiplication is *distributive* with respect to addition. The proof is elementary, and is left to the student. It is customary to omit $\cdot$ in the product of two complex numbers.

Two elementary consequences of the definition of **C** are the following:

1. Given two complex numbers $\alpha = (a,b)$ and $\beta = (c,d)$, the equation $\alpha + z = \beta$ has the unique solution $z = (c - a, d - b)$. The complex number z is called the *difference* of α from β and is denoted by $\beta - \alpha$.
2. Given two complex numbers $\alpha = (a,b)$ and $\beta = (c,d)$ with $\alpha \neq (0,0)$, the equation $\alpha z = \beta$ has the unique solution

$$z = \left(\frac{ac + bd}{a^2 + b^2}, \frac{ad - bc}{a^2 + b^2} \right)$$

This complex number z is called the quotient of β over α, denoted by β/α. Note that $\beta/\alpha = \beta\alpha^{-1}$.

Example 1.1 The quadratic equation $z^2 - (4,0)z + (5,0) = (0,0)$ has no root in the field of real numbers. However, it has the two complex roots $(2,1)$ and $(2,-1)$.

Proof. Let us show that $(2,-1)$ is a root. In fact,

$$
\begin{aligned}
(2,-1)^2 - (4,0)(2,-1) + (5,0) &= (2,-1)(2,-1) - (4,0)(2,-1) + (5,0) \\
&= (3,-4) - (8,-4) + (5,0) \\
&= (0,0)
\end{aligned}
$$

The field **R** of real numbers is *isomorphically embedded* into the field **C** of complex numbers. That is, the real numbers can be put in a one-to-one correspondence with complex numbers of the form $(a,0)$, where a is real, by the rule

$$a \leftrightarrow (a,0) \qquad \text{for } a \in \mathbf{R}$$

and this rule preserves the algebraic structures. This is seen from the relations

$$(a,0) + (b,0) = (a + b, 0) \leftrightarrow a + b$$

$$(a,0)(b,0) = (ab,0) \leftrightarrow ab$$

Because of this isomorphism, for algebraic purposes the complex number $(a,0)$ can be identified with the real number a. Therefore we shall write $a = (a,0)$, and we shall consider the real numbers as a subset of the complex numbers. Note in particular that $0 = (0,0)$.

If we introduce the symbol i to represent the complex number $(0,1)$, we see that

$$i^2 = (0,1)(0,1) = (-1,0) = -(1,0) = -1$$

which agrees with Euler's definition of the symbol i. (See Euler's biography.) Moreover,

$$(a,b) = (a,0) + (0,b) = (a,0) + (0,1)(b,0) = a + ib$$

Thus (a,b) may be denoted by $a + ib$ with the understanding that a is $(a,0)$, $i = (0,1)$, $b = (b,0)$, and $a + ib$ is the sum of a plus (i times b). In the sequel, we prefer to write complex numbers in the form $a + ib$. When we write $a + ib$ without further explanation, it is understood that a and b are real numbers. In the complex number $z = a + ib$, a is called the *real part* of z and b the *imaginary part* of z. We shall use the notation $a = \operatorname{Re} z$ and $b = \operatorname{Im} z$. If the real part of a complex number is zero, the number is called *pure-imaginary*.

In performing operations with complex numbers of the form $a + ib$, we can proceed as in the algebra of real numbers, replacing i^2 by -1 when it occurs. Thus we have the following expressions.

Addition:

$$(a + ib) + (c + id) = a + ib + c + id = (a + c) + i(b + d)$$

Multiplication:

$$(a + ib)(c + id) = ac + iad + ibc + i^2bd = (ac - bd) + i(ad + bc)$$

Subtraction:

$$(a + ib) - (c + id) = a + ib - c - id = (a - c) + i(b - d)$$

Division:

For $c + id \neq 0$ (or equivalently, $c^2 + d^2 \neq 0$),

$$\frac{a + ib}{c + id} = \frac{a + ib}{c + id} \cdot \frac{c - id}{c - id} = \frac{ac - iad + ibc - i^2bd}{c^2 - icd + idc - i^2d^2}$$

$$= \frac{(ac + bd) + i(bc - ad)}{c^2 + d^2} = \frac{ac + bd}{c^2 + d^2} + i\,\frac{bc - ad}{c^2 + d^2}$$

Example 1.2 It follows from the above operations that

$$\frac{(-2 + 5i)(1 + 3i)}{2 + 3i} - \left(\frac{2}{13} - \frac{3i}{13}\right) = \frac{-17 - i}{2 + 3i} - \frac{2 - 3i}{13}$$

$$= \frac{(-17 - i)(2 - 3i)}{(2 + 3i)(2 - 3i)} - \frac{2 - 3i}{13}$$

$$= \frac{-37 + 49i}{13} - \frac{2 - 3i}{13}$$

$$= -3 + 4i$$

(Sometimes it is convenient to write $a + bi$ instead of $a + ib$.)

LEONHARD EULER

Leonhard Euler was born at Basle, Switzerland, on April 15, 1707. His father, Paul Euler, was a Calvinist pastor with a strong background in mathematics, which he acquired as a pupil of the famous Jakob Bernoulli. Paul's hopes that his son would pursue a theological career did not materialize, partly because he made the unregrettable mistake of teaching Leonhard mathematics.

Euler, in fact, turned out to be one of the greatest scientists that Switzerland has ever produced and the most prolific mathematician in history. The academies of science of both Prussia and Russia could not publish as fast as he wrote. It is impossible in a brief summary to do justice to the variety and importance of his work. His interests ranged over a broad spectrum from elementary geometry to the analytic treatment of trigonometric functions, the calculus of variations, and number theory. He wrote several classic textbooks and also proved a number of new results that now bear his name. He is considered one of the founders of modern analysis, the field in which he made his greatest contribution.

Euler was fortunate in finding patrons in both Catherine I and Frederick the Great, who appointed him to academic positions that enabled him to pursue his mathematical studies without interruption. Even his busy family life (he had 13 children) and his total blindness during the last 17 years of his life did not stop his flood of publications.

Euler died in St. Petersburg, Russia, in 1783 while playing with one of his grandchildren during a tea break. He was 76 years old.

Powers of complex numbers with integral exponents are defined in the same way that they are defined for real numbers. Thus we define

$$a^1 = a \qquad \text{for every } a \in \mathbf{C}$$

and inductively we define

$$a^n = a^{n-1} \cdot a \qquad \text{for every } a \in \mathbf{C} \text{ and every integer } n \geq 2$$

$$a^{-n} = \frac{1}{a^n} \qquad \text{for every } a \in \mathbf{C}, a \neq 0, \text{ and every integer } n \geq 1$$

$$a^0 = 1 \qquad \text{for every } a \in \mathbf{C}, a \neq 0$$

Polynomials and rational functions (quotient of two polynomials) in $\mathbf{C}$ are also defined as in $\mathbf{R}$. The following properties of powers, familiar to us from real numbers, are again valid in $\mathbf{C}$:

$$a^n a^m = a^{n+m} \qquad (a^n)^m = a^{nm} \qquad (ab)^n = a^n b^n$$

(where a and b cannot be equal to 0 if their exponents are negative). The expression 0^0, as in the case of real numbers, is not meaningful for complex numbers. If n is a positive integer, one can prove by induction that Newton's formula (Isaac Newton, 1642–1727, English)

$$(a + b)^n = \sum_{k=0}^{n} \binom{n}{k} a^{n-k} b^k$$

is also valid for all complex numbers a and b, where

$$\binom{n}{0} = 1 \qquad \text{and} \qquad \binom{n}{k} = \frac{n(n-1)\cdots(n-k+1)}{k!} \qquad \text{for } 1 \leq k \leq n$$

Since $i^2 = -1$, we have for any integer $n = 0, 1, 2, \ldots$

$$i^{4n} = (i^4)^n = 1^n = 1 \qquad i^{4n+1} = i^{4n} \cdot i = i$$

$$i^{4n+2} = i^{4n} \cdot i^2 = -1 \qquad i^{4n+3} = i^{4n} \cdot i^3 = -i$$

Example 1.3 Show that $(1 + i)^{100} = -2^{50}$.

Proof. In fact,

$$(1 + i)^{100} = [(1 + i)^2]^{50} = (2i)^{50} = 2^{50} \cdot i^{50} = 2^{50}(-1) = -2^{50}$$

EXERCISES

1.2.1 Show that $(1,2)$ and $(1,-2)$ are roots of the equation

$$z^2 - (2,0)z + (5,0) = (0,0)$$

1.2.2 Show that

$$\frac{(-1 - 3i)(1 - 2i)}{2 + i} + 3i = -3 + 4i$$

1.2.3 Show that $(1 - i\sqrt{3})^{-10} = -2^{-11}(1 + i\sqrt{3})$.

1.2.4 For any complex numbers z_1 and z_2, show that

    a. $\operatorname{Re}(z_1 + z_2) = \operatorname{Re} z_1 + \operatorname{Re} z_2$

    b. $\operatorname{Im}(z_1 + z_2) = \operatorname{Im} z_1 + \operatorname{Im} z_2$

    c. $\operatorname{Re} z_1 z_2 = \operatorname{Re} z_1 \operatorname{Re} z_2 - \operatorname{Im} z_1 \operatorname{Im} z_2$

    d. $\operatorname{Im} z_1 z_2 = \operatorname{Re} z_1 \operatorname{Im} z_2 + \operatorname{Im} z_1 \operatorname{Re} z_2$

ISAAC NEWTON

Isaac Newton was born in Woolsthorpe, England, on Christmas Day, 1642, about two months after the death of his father. His mother remarried when Isaac was three, and left him in the care of his maternal grandmother. When Isaac was 14, his mother was again widowed and called Isaac to her to help run the family farm.

The young Newton was far more interested in mathematics than in plowing, however, and when he was 18 an uncle persuaded his mother to allow her son to prepare for Cambridge. During his year of preparation he became engaged to a Miss Storey. They parted affectionately when he went to Cambridge, but the marriage never took place. Newton remained a bachelor all his days.

At Cambridge, Newton studied elementary mathematics. However the Great Plague of 1664–1665 forced a closing of the university, and Newton retired to his home in Woolsthorpe. There, in an astonishing burst of mathematical creativity, he invented the calculus (including the binomial theorem), discovered the visible spectrum in sunlight, and formulated the law of universal gravitation. All this before he was 25 years old, and in a period of only 18 months as well!

At Cambridge once again, Newton was made Lucasian Professor of Mathematics when his teacher, Isaac Barrow, resigned the chair in favor of his pupil. Newton was 26 at the time.

Newton's enduring fame is based on his *Philosphiae Naturalis Principia Mathematica*, which was published in 1687. The *Principia* showed how a unified mathematical law could account for motions of the solar system, tides, and other gravitational phenomena. It guided the course of physics and astronomy for well over 200 years.

In his later years, Newton was drawn into politics, and in 1696 was made Warden of the Mint. He served in that capacity and as Master of the Mint until his death on March 20, 1727, at the age of 84.

1.2.5 Reduce to the form $a + ib$.

    a. $i^{26} - 3i^7 + i^6(1 - i^3) - (-i)^{18}$

    b. $\dfrac{(2 + 3i)(-1 + 2i)}{2 + i} - \dfrac{1 - i}{1 - 2i}$

1.2.6 Show that $1 \pm i$ are roots of the polynomial equation $z^4 - 2z^3 + 3z^2 - 2z + 2 = 0$, and then factor the polynomial in the field of real numbers.

1.2.7 Given that $z = x + iy$, show that z^4 is real if and only if $xy = 0$ or $|x| = |y|$, and z^4 is pure-imaginary if and only if $x = \pm(1 \pm \sqrt{2})y$.

1.3
COMPLEX CONJUGATES AND
ABSOLUTE VALUES

    The *complex conjugate* (or simply the *conjugate*) of the complex number $z = x + iy$ is denoted by $\bar{z}$ and is defined by $\bar{z} = x - iy$. For example, $\overline{6 - 7i} = 6 + 7i$. From the definition of conjugate it follows that

$$\overline{z_1 + z_2} = \bar{z}_1 + \bar{z}_2 \tag{1.1}$$

and by induction

$$\overline{z_1 + z_2 + \cdots + z_n} = \bar{z}_1 + \bar{z}_2 + \cdots + \bar{z}_n \tag{1.1a}$$

$$\overline{z_1 z_2} = \bar{z}_1 \bar{z}_2 \tag{1.2}$$

and by induction

$$\overline{z_1 z_2 \cdots z_n} = \bar{z}_1 \bar{z}_2 \cdots \bar{z}_n \tag{1.2a}$$

$$\overline{z_1 - z_2} = \bar{z}_1 - \bar{z}_2 \tag{1.3}$$

$$\overline{\left(\frac{z_1}{z_2}\right)} = \frac{\bar{z}_1}{\bar{z}_2} \qquad z_2 \neq 0 \tag{1.4}$$

That is, the conjugate of the sum, product, difference, or quotient of two complex numbers is equal, respectively, to the sum, product, difference, or quotient of their conjugates.

$$\overline{\bar{z}} = z \tag{1.5}$$

$$\text{Re } z = \frac{z + \bar{z}}{2} \qquad \text{and} \qquad \text{Im } z = \frac{z - \bar{z}}{2i} \tag{1.6}$$

From Properties (1.1a), (1.2a), (1.3), and (1.4), it follows that if $Q(z_1, z_2, \ldots, z_n)$ is a rational function of $z_1, z_2, \ldots, z_n$ with real coefficients, then

$$\overline{Q(z_1, z_2, \ldots, z_n)} = Q(\bar{z}_1, \bar{z}_2, \ldots, \bar{z}_n)$$

Example 1.4 If z_0 is a root of the polynomial equation

$$P(z) = a_n z^n + a_{n-1} z^{n-1} + \cdots + a_1 z + a_0 = 0$$

where all the coefficients $a_0, a_1, \ldots, a_n$ are real, then $\bar{z}_0$ is also a root of $P(z) = 0$.

Proof. In fact, $P(z_0) = 0$, and therefore

$$P(\bar{z}_0) = \overline{P(z_0)} = \bar{0} = 0$$

The *absolute value* or *modulus* of the complex number $z = x + iy$ is denoted by $|z|$ and is defined to be the nonnegative square root of $x^2 + y^2$. Thus $|z|^2 = x^2 + y^2 = z\bar{z}$. For example, $|3 - 4i| = \sqrt{3^2 + (-4)^2} = \sqrt{9 + 16} = 5$. From the definition of absolute value, it follows that:

$$|z| = 0 \quad \text{if and only if} \quad z = 0 \tag{1.7}$$

$$|z| = |\bar{z}| \tag{1.8}$$

$$z^{-1} = \frac{\bar{z}}{|z|^2} \qquad z \neq 0 \tag{1.9}$$

$$|\operatorname{Re} z| \leq |z| \quad \text{and} \quad |\operatorname{Im} z| \leq |z| \tag{1.10}$$

$$|a| = |(a,0)| \qquad a \in \mathbf{R} \tag{1.11}$$

and so the sense of length in $\mathbf{R}$ is preserved by the embedding $a \leftrightarrow (a,0)$,

$$|z_1 z_2| = |z_1| \, |z_2| \tag{1.12}$$

To prove (1.12), observe that

$$\begin{aligned} |z_1 z_2|^2 &= (z_1 z_2)(\overline{z_1 z_2}) = (z_1 z_2)(\bar{z}_1 \, \bar{z}_2) \\ &= (z_1 \bar{z}_1)(z_2 \bar{z}_2) = |z_1|^2 |z_2|^2 \end{aligned}$$

from which (1.12) follows, since the absolute value of a complex number is nonnegative. By induction, we have

$$|z_1 z_2 \cdots z_n| = |z_1| \, |z_2| \cdots |z_n| \tag{1.12a}$$

That is, the absolute value of a finite product of complex numbers is the product of absolute values of the factors:

$$\left| \frac{z_1}{z_2} \right| = \frac{|z_1|}{|z_2|} \qquad z_2 \neq 0 \tag{1.13}$$

That is, the absolute value of the quotient of two complex numbers is the quotient of the absolute values.

$$|z_1 + z_2| \leq |z_1| + |z_2| \qquad \textit{(triangle inequality)} \tag{1.14}$$

To prove (1.14), observe that [in view of (1.10) and (1.8)]

$$|z_1 + z_2|^2 = (z_1 + z_2)(\overline{z_1 + z_2}) = (z_1 + z_2)(\bar{z}_1 + \bar{z}_2)$$
$$= z_1\bar{z}_1 + z_1\bar{z}_2 + \bar{z}_1 z_2 + z_2\bar{z}_2 = |z_1|^2 + z_1\bar{z}_2 + \overline{z_1\bar{z}_2} + |z_2|^2$$
$$= |z_1|^2 + 2\,\mathrm{Re}\,(z_1\bar{z}_2) + |z_2|^2 \leq |z_1|^2 + 2|z_1\bar{z}_2| + |z_2|^2$$
$$= |z_1|^2 + 2|z_1|\,|z_2| + |z_2|^2 = (|z_1| + |z_2|)^2$$

from which the triangle inequality follows. By induction, one can establish that

$$|z_1 + z_2 + \cdots + z_n| \leq |z_1| + |z_2| + \cdots + |z_n| \qquad (1.14a)$$

That is, the absolute value of the sum of a finite number of complex numbers cannot exceed the sum of their absolute values.

$$\big||z_1| - |z_2|\big| \leq |z_1 - z_2| \qquad (1.15)$$

In fact, as in Eq. (1.14a),

$$|z_1 - z_2|^2 = (z_1 - z_2)(\overline{z_1 - z_2}) = (z_1 - z_2)(\bar{z}_1 - \bar{z}_2)$$
$$= |z_1|^2 - 2\,\mathrm{Re}\,(z_1\bar{z}_2) + |z_2|^2$$
$$\geq |z_1|^2 - 2|z_1|\,|z_2| + |z_2|^2$$
$$= (|z_1| - |z_2|)^2 = \big||z_1| - |z_2|\big|^2$$

from which Eq. (1.15) follows.

Example 1.5 Let a_j and b_j ($j = 1, 2, \ldots, n$) be complex numbers. Prove *Lagrange's identity* (Joseph Louis Lagrange, 1736–1813, French):

$$\left|\sum_{j=1}^{n} a_j b_j\right|^2 = \sum_{j=1}^{n} |a_j|^2 \sum_{j=1}^{n} |b_j|^2 - \sum_{1 \leq j < k \leq n} |a_j\bar{b}_k - a_k\bar{b}_j|^2$$

Proof. We have

$$\left|\sum_{j=1}^{n} a_j b_j\right|^2 = \sum_{j=1}^{n} a_j b_j \sum_{j=1}^{n} \overline{a_j b_j}$$

$$= \sum_{j=1}^{n} |a_j b_j|^2 + \sum_{1 \leq j < k \leq n} (a_j b_j \overline{a_k b_k} + a_k b_k \overline{a_j b_j})$$

Also,

$$\sum_{j=1}^{n} |a_j|^2 \sum_{j=1}^{n} |b_j|^2 = \sum_{j=1}^{n} a_j\bar{a}_j \sum_{j=1}^{n} b_j\bar{b}_j$$

$$= \sum_{j=1}^{n} |a_j b_j|^2 + \sum_{1 \leq j < k \leq n} (a_j b_k \overline{a_j b_k} + a_k b_j \overline{a_k b_j})$$

Lagrange's inequality follows after solving the second identity with respect to the term $\sum_{j=1}^{n} |a_j b_j|^2$ and substituting its value in the first.

Example 1.6 Let a_j and b_j $(j = 1, 2, \ldots, n)$ be complex numbers. Prove *Cauchy's inequality*:

$$\left| \sum_{j=1}^{n} a_j b_j \right|^2 \leq \sum_{j=1}^{n} |a_j|^2 \sum_{j=1}^{n} |b_j|^2$$

Proof. It follows immediately from Lagrange's identity.

EXERCISES

1.3.1 Prove that $z = \bar{z}$ if and only if z is a real number and that $z = -\bar{z}$ if and only if z is pure-imaginary.

1.3.2 Prove that $z^2 = (\bar{z})^2$ if and only if z is either real or pure-imaginary.

1.3.3 Find the roots of $z^4 - 2z^3 + 6z^2 - 8z + 8 = 0$ given that $1 - i$ is one root.

1.3.4 Show that

$$\left| \frac{(3 - 4i)(2 + i)}{(2 - 4i)(6 + 8i)} \right| = \frac{1}{4}$$

1.3.5 For any complex numbers z_1 and z_2, prove the parallelogram law:

$$|z_1 + z_2|^2 + |z_1 - z_2|^2 = 2(|z_1|^2 + |z_2|^2)$$

1.3.6 Prove that $|z| \leq |\text{Re } z| + |\text{Im } z| \leq \sqrt{2}\, |z|$.

1.3.7 Given that $|a| \leq 1$, show that $|z| \leq 1$ is equivalent to

$$\left| \frac{z - a}{1 - \bar{a}z} \right| \leq 1$$

and that the equality holds if and only if $|z| = 1$.

1.3.8 Suppose that $|z_j| < 1, \lambda_j \geq 0$ for $j = 1, 2, \ldots, n$, and $\lambda_1 + \lambda_2 + \cdots + \lambda_n = 1$. Show that $|\lambda_1 z_1 + \cdots + \lambda_n z_n| < 1$.

1.3.9 Give an independent proof of Cauchy's inequality (Example 1.6) by minimizing the real-valued function:

$$F(\lambda) = \sum_{j=1}^{n} (|a_j| - \lambda|b_j|)^2 \qquad \lambda \in \mathbf{R}$$

1.3.10 Determine when equality holds in Eq. (1.15). That is, for what $z_1, z_2 \in \mathbf{C}$ do we have $\big| |z_1| - |z_2| \big| = |z_1 - z_2|$?

1.3.11 Determine when equality holds in Cauchy's inequality.

1.4

THE GEOMETRIC REPRESENTATION OF COMPLEX NUMBERS

The theory of complex variables is a branch of mathematical analysis, not a branch of geometry. Therefore all the conclusions in complex variables should be derived from the axioms of real numbers, not from the axioms of geometry. However, the geometric language and ideas are very suggestive and are often used with great advantage to interpret analytical concepts. For example, in 1799 Gauss interpreted what was, at that time, the artificial concept of the complex number $x + iy$ as representing the point (x, y) in a rectangular coordinate system for the plane $\mathbf{R} \times \mathbf{R}$.

This representation (which we have accepted) established a one-to-one correspondence between the set of complex numbers and the plane $\mathbf{R} \times \mathbf{R}$ given by $x + iy \leftrightarrow (x, y)$. It is traditional to write $z = x + iy$ and to call $\mathbf{R} \times \mathbf{R}$ the *complex plane*, or *z plane*. The x and y axes are called the *real* and *imaginary axes*, respectively. Instead of considering the complex number $z = x + iy$ as the ordered pair (x, y), it will frequently aid our intuition to consider the directed line segment, or vector, extended from the origin to (x, y) as representing z. As usual, any vector $\mathbf{AB}$ with horizontal coordinate x and vertical coordinate y will also be taken to represent $z = x + iy$ (that is, we identify all vectors which can be obtained from each other by parallel displacement). See Fig. 1.1.

Addition of complex numbers is simply vector addition according to the parallelogram law, as shown in Fig. 1.2. Thus to add the complex numbers z_1 and z_2, complete the parallelogram with adjacent sides z_1 and z_2. Its diagonal from the origin represents $z_1 + z_2$.

Fig. 1.1

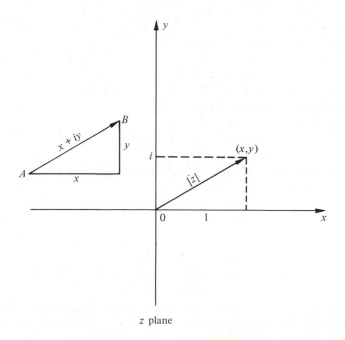

z plane

The *difference* $z_1 - z_2$ can be found by adding $-z_2$ to z_1 and, as we see from Fig. 1.3, is the vector from z_2 to z_1.

If $z = x + iy$, then $|z| = +\sqrt{x^2 + y^2}$. Hence $|z|$ is the length of the vector **z**. (See Fig. 1.1.)

If $z_1 = x_1 + iy_1$ and $z_2 = x_2 + iy_2$, then

$$|z_1 - z_2| = +\sqrt{(x_1 - x_2)^2 + (y_1 - y_2)^2}$$

Thus $|z_1 - z_2|$ is the distance from z_1 to z_2, that is, the length of the vector from z_2 to z_1. This observation enables us to give analytic representations of geometric concepts. Thus the set $\{z \in \mathbf{C} : |z - z_0| = r\}$ represents a *circle* in the z plane centered at z_0 with radius r. Similarly, the set

$$\{z \in \mathbf{C} : |z - z_0| < r\}$$

represents an *open disk* in the z plane, centered at z_0 with radius r, and

$$\{z \in \mathbf{C} : |z - z_0| \leq r\}$$

represents a *closed disk* in the z plane, centered at z_0 with radius r. Notice that although *circle* and *disk* are geometric concepts, the above definitions are analytic.

Fig. 1.2

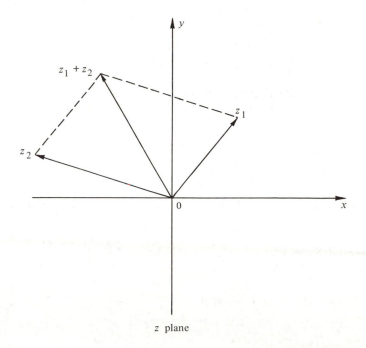

z plane

Fig. 1.3

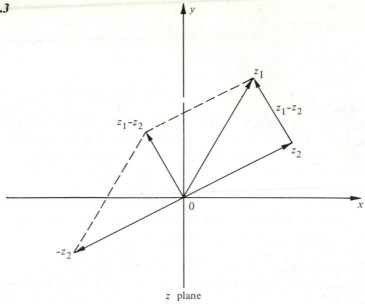

z plane

Example 1.7 Let z_1, z_2, and z_3 be three complex numbers such that

$$|z_1| = |z_2| = |z_3| = 1 \quad \text{and} \quad z_1 + z_2 + z_3 = 0$$

Then z_1, z_2, z_3 are the vertices of an equilateral triangle inscribed in the unit circle $\{z \in \mathbf{C}: |z| = 1\}$.

Proof. Clearly, it suffices to show that

$$|z_1 - z_2|^2 = |z_2 - z_3|^2 = |z_3 - z_1|^2$$

in order to prove the assertion. We have

$$|z_1 - z_2|^2 = |2z_1 + z_3|^2 = (2z_1 + z_3)(2\bar{z}_1 + \bar{z}_3)$$
$$= 5 + 2(z_1\bar{z}_3 + \bar{z}_1 z_3)$$

Similarly,

$$|z_2 - z_3|^2 = |-z_1 - 2z_3|^2 = |z_1 + 2z_3|^2$$
$$= 5 + 2(z_1\bar{z}_3 + \bar{z}_1 z_3) = |z_1 - z_2|^2$$

A similar argument shows that $|z_2 - z_3|^2 = |z_3 - z_1|^2$; the proof is complete.

Example 1.8 Show that all the roots of the equation $z^3 + 2z + 4 = 0$ lie outside the unit circle.

Proof. In fact, for $|z| \leq 1$, we have

$$|z^3 + 2z + 4| \geq 4 - |z|^3 - 2|z| \geq 4 - 1 - 2 = 1 > 0$$

which proves our result.

In order to obtain a geometric interpretation of the *product* and *quotient* of two complex numbers, we introduce polar coordinates (r,θ). See Fig. 1.4. In view of the relations

$$x = r \cos \theta \qquad y = r \sin \theta$$

we have

$$z = x + iy = r(\cos \theta + i \sin \theta)$$

This is the *polar form* of the complex number z. Here $r \geq 0$ is the modulus $|z|$ of z. The coordinate θ is undefined for $z = 0$, but if $z \neq 0$, then θ is any real number such that

$$\cos \theta = \frac{x}{|z|} \quad \text{and} \quad \sin \theta = \frac{y}{|z|} \tag{1.16}$$

Because of the periodicity of the sine and cosine, θ is determined only to within some multiple of 2π. The coordinate θ is called the *argument* of z and is denoted by arg z. Thus if $\theta \in \mathbf{R}$ satisfies (1.16), then

$$\arg z = \{\theta + 2\pi n : n = 0, \pm 1, \pm 2, \ldots\} \tag{1.17}$$

If $\theta + 2\pi n$ is any value of arg z, we shall write arg $z = \theta \bmod 2\pi$ (where $a = b \bmod 2\pi$ means $a - b = 2\pi k$ for some integer k). Instead of (1.17) we may also write

$$\arg z = \theta + 2\pi n \qquad n = 0, \pm 1, \pm 2, \ldots$$

Such a function is called *multiple-valued*.

The choice of arg z such that

$$-\pi < \arg z \leq \pi \tag{1.18}$$

Fig. 1.4

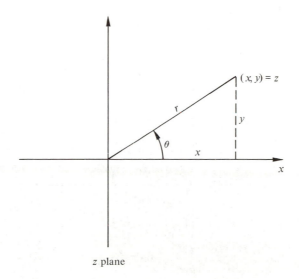

z plane

is referred to as the *principal value* of the argument of z and is denoted by Arg z. The choice of the interval $(-\pi, \pi]$ in (1.18) is arbitrary; we could have used any interval of the form $(\theta_0, \theta_0 + 2\pi]$ or $[\theta_0, \theta_0 + 2\pi)$ in defining the *single-valued* function Arg.

Let $z_1 = |z_1|(\cos \theta_1 + i \sin \theta_1)$ and $z_2 = |z_2|(\cos \theta_2 + i \sin \theta_2)$ be two complex numbers in polar form. Then, using the addition theorems of sine and cosine (which one can establish analytically), we obtain

$$
\begin{aligned}
z_1 z_2 &= |z_1|\,|z_2|[(\cos \theta_1 \cos \theta_2 - \sin \theta_1 \sin \theta_2) \\
&\quad + i(\sin \theta_1 \cos \theta_2 + \cos \theta_1 \sin \theta_2)] \\
&= |z_1 z_2|[\cos (\theta_1 + \theta_2) + i \sin (\theta_1 + \theta_2)]
\end{aligned}
$$

See Fig. 1.5.

Thus the modulus of the product of two complex numbers is the product of their moduli, and the argument is the sum of the two arguments (up to a multiple of 2π). That is,

$$
\arg z_1 z_2 = \arg z_1 + \arg z_2 \qquad \mathrm{mod}\ 2\pi
$$

If $z_2 \neq 0$, we have (after some manipulations)

$$
\frac{z_1}{z_2} = \frac{|z_1|(\cos \theta_1 + i \sin \theta_1)}{|z_2|(\cos \theta_2 + i \sin \theta_2)}
$$

$$
= \left|\frac{z_1}{z_2}\right| [\cos (\theta_1 - \theta_2) + i \sin (\theta_1 - \theta_2)]
$$

Fig. 1.5

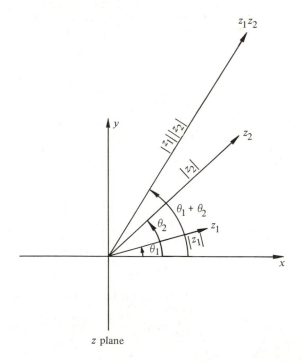

z plane

Thus the modulus of the quotient of two complex numbers is the quotient of their moduli, and the argument is the difference of the two arguments (up to a multiple of 2π). That is,

$$\arg \frac{z_1}{z_2} = \arg z_1 - \arg z_2 \qquad \text{mod } 2\pi$$

By induction, we easily establish that

$$z_1 z_2 \cdots z_n = |z_1 z_2 \cdots z_n|[\cos(\theta_1 + \theta_2 + \cdots + \theta_n) + i \sin(\theta_1 + \theta_2 + \cdots + \theta_n)]$$

In particular, if $z = |z|(\cos\theta + i\sin\theta)$, then $z^n = |z|^n(\cos n\theta + i\sin n\theta)$. Hence the following identity of Abraham De Moivre (1667–1754, French) holds:

$$(\cos\theta + i\sin\theta)^n = \cos n\theta + i\sin n\theta \qquad (1.19)$$

ABRAHAM DE MOIVRE

Abraham De Moivre was born of a French Huguenot family at Vitry in Champagne on May 20, 1667. His father was a surgeon. After the revocation of the Edict of Nantes and the expulsion of the Huguenots, De Moivre settled in London in 1685. In England he eked out a living by giving private lessons in mathematics and games of chance. He never married.

De Moivre became personally acquainted with Sir Isaac Newton and Edmund Halley. He studied Newton's *Principia* very carefully and soon became recognized as a great mathematician. In 1697 he was elected to the Royal Society and in 1712 he served as one of the commissioners to decide the rival claims of Newton and Leibniz regarding priority in the discovery of the infinitesimal calculus.

De Moivre is considered to be one of the founders of probability theory, to which he contributed significantly. He also contributed to the calculus and the theory of equations. The so-called "De Moivre theorem" is used in papers that he published in 1707 and 1730.

In spite of the fact that De Moivre was well recognized as a mathematician, he never secured a university position, perhaps because he was not British by birth.

He died, almost blind, on November 27, 1754, at the age of 87.

This is called *De Moivre's theorem*. It is also true for negative integers, for if $z = \cos\theta + i\sin\theta$, then

$$z^{-1} = (\cos\theta + i\sin\theta)^{-1} = \cos(-\theta) + i\sin(-\theta)$$
$$z^{-n} = (z^{-1})^n$$

Thus De Moivre's identity (1.19) is valid for any integer n.

Example 1.9 Let

$$z = \frac{(1-i)^8}{(\sqrt{3}+i)^5}$$

Write z in polar form and find its principal argument.

Solution. We have

$$1 - i = \sqrt{2}[\cos(-\pi/4) + i\sin(-\pi/4)]$$

and

$$\sqrt{3} + i = 2[\cos(\pi/6) + i\sin(\pi/6)]$$

Thus

$$(1-i)^8 = 16[\cos(-2\pi) + i\sin(-2\pi)]$$

and

$$(\sqrt{3}+i)^5 = 32\left(\cos\frac{5\pi}{6} + i\sin\frac{5\pi}{6}\right)$$

Hence

$$z = \frac{16[\cos(-2\pi) + i\sin(-2\pi)]}{32[\cos(5\pi/6) + i\sin(5\pi/6)]}$$

$$= \frac{1}{2}\left[\cos\left(-\frac{17\pi}{6}\right) + i\sin\left(-\frac{17\pi}{6}\right)\right]$$

and $\operatorname{Arg} z = -5\pi/6$.

Example 1.10 Show that two triangles with vertices z_1, z_2, z_3 and ζ_1, ζ_2, ζ_3 are similar if and only if

$$\begin{vmatrix} 1 & 1 & 1 \\ z_1 & z_2 & z_3 \\ \zeta_1 & \zeta_2 & \zeta_3 \end{vmatrix} = 0 \tag{1.20}$$

Proof. Recall that two triangles ABC and $A'B'C'$ are similar if and only if

$$\angle A = \angle A' \quad \text{and} \quad \frac{AC}{AB} = \frac{A'C'}{A'B'} \tag{1.21}$$

Let $A = z_1$, $B = z_2$, $C = z_3$, $A' = \zeta_1$, $B' = \zeta_2$, and $C' = \zeta_3$. Clearly (1.21) is equivalent to

$$\operatorname{Arg}\frac{z_3 - z_1}{z_2 - z_1} = \operatorname{Arg}\frac{\zeta_3 - \zeta_1}{\zeta_2 - \zeta_1}$$

and

$$\left| \frac{z_3 - z_1}{z_2 - z_1} \right| = \left| \frac{\zeta_3 - \zeta_1}{\zeta_2 - \zeta_1} \right|$$

which in turn is equivalent to

$$\frac{z_3 - z_1}{z_2 - z_1} = \frac{\zeta_3 - \zeta_1}{\zeta_2 - \zeta_1}$$

which is equivalent to Eq. (1.20).

EXERCISES

1.4.1 Compute graphically
    a. $(2 - 3i) + (-1 + 2i)$ *b.* $(2 + 3i) - (3 - 2i)$
    c. $(1 + i)(\sqrt{3} - i)$ *d.* $(1 + i\sqrt{3})/(1 - i\sqrt{3})$

1.4.2 Write in polar form the complex number $z = (1 - i\sqrt{3})^7/(\sqrt{3} - i)^3$ and compute its principal argument.

1.4.3 Show graphically that the four points $z, \bar{z}, -z, -\bar{z}$ are the vertices of a rectangle which is symmetric with respect to both axes.

1.4.4 Find in terms of the "complex variable" z the equation for
    a. a straight line through the points $2 - 3i$ and $-1 + 2i$
    b. a circle of radius 3 centered at $-2 + 5i$
    c. an ellipse with foci at $3i$ and $-3i$ with major axis of length 8

1.4.5 Show that the equation of the straight-line segment joining z_1 and z_2 is given by

$$z(t) = z_1 + t(z_2 - z_1) \qquad 0 \le t \le 1$$

1.4.6 Show that the triangle with vertices z_1, z_2, z_3 is equilateral if and only if

$$z_1^2 + z_2^2 + z_3^2 = z_1 z_2 + z_2 z_3 + z_3 z_1$$

1.4.7 Interpret graphically the transformation $w = 2iz + 1$.

1.4.8 If z_1 and z_2 are two fixed points, find the locus of the points z for which

    a. $\left| \dfrac{z - z_1}{z - z_2} \right| = $ constant   *b.* $\text{Arg}\, \dfrac{z - z_1}{z - z_2} = $ constant

1.4.9 Show that the sum of the angles of a triangle is π.

1.4.10 Use De Moivre's theorem to show that
    a. $\sin 3\theta = 3 \cos^2 \theta \sin \theta - \sin^3 \theta$
    b. $\cos 3\theta = \cos^3 \theta - 3 \cos \theta \sin^2 \theta$
    c. $\cos 8\theta + 28 \cos 4\theta + 35 = 64(\cos^8 \theta + \sin^8 \theta)$

1.4.11 The distance between two complex numbers z_1 and z_2 is defined by $d(z_1, z_2) = |z_1 - z_2|$. Show that d defines a *metric* on **C**; that is, for any $z_1, z_2, z_3 \in \mathbf{C}$,
    a. $d(z_1, z_2) \ge 0$ and $d(z_1, z_2) = 0$ if and only if $z_1 = z_2$
    b. $d(z_1, z_2) = d(z_2, z_1)$
    c. $d(z_1, z_3) \le d(z_1, z_2) + d(z_2, z_3)$ (triangle inequality)
    [In Theorem 2.1 we shall prove that $(\mathbf{C}, d)$ is a *complete* metric space.]

1.5
ROOTS OF COMPLEX NUMBERS

Let a be a nonzero complex number and let n be a positive integer. The nth roots of a are by definition the set $a^{1/n} = \{z \in \mathbf{C}: z^n = a\}$. Thus if z is an nth root of a, then

$$z^n = a \tag{1.22}$$

Let $a = |a|(\cos \theta + i \sin \theta)$ and $z = |z|(\cos \phi + i \sin \phi)$ be the polar forms of a and z, respectively. Then from (1.22) and De Moivre's theorem, we have

$$|z|^n(\cos n\phi + i \sin n\phi) = |a|(\cos \theta + i \sin \theta) \tag{1.23}$$

Thus, if $\sqrt[n]{|a|}$ denotes the real positive nth root of $|a|$, equating real and imaginary parts in Eq. (1.23) yields

$$|z| = \sqrt[n]{|a|} \quad \text{and} \quad \phi = \frac{\theta + 2\pi k}{n} \quad \text{for some integer } k$$

Because of the periodicity of sine and cosine, we can easily see that for $k = n + m$ we get the same nth root of a as when $k = m$. Hence

$$\{z \in \mathbf{C}: z^n = a\}$$

$$= \left\{ \sqrt[n]{|a|} \left(\cos \frac{\theta + 2\pi k}{n} + i \sin \frac{\theta + 2\pi k}{n} \right): \quad k = 0, 1, 2, \ldots, n - 1 \right\}$$

We write this symbolically as

$$z = a^{1/n} = \sqrt[n]{|a|} \left(\cos \frac{\theta + 2\pi k}{n} + i \sin \frac{\theta + 2\pi k}{n} \right)$$

$$k = 0, 1, 2, \ldots, n - 1 \tag{1.24}$$

Therefore a has exactly n nth roots.

It is easily seen that (1.24) is also valid for any negative integer. Similarly, we define $a^{m/n} = (a^m)^{1/n}$. Note that

$$a^{m/n} = \sqrt[n]{|a|^m} \left(\cos \frac{m\theta + 2\pi k}{n} + i \sin \frac{m\theta + 2\pi k}{n} \right) \quad k = 0, \ldots, n - 1$$

Hence $a^{m/n}$ has n elements. Note also that

$$(a^{1/n})^m = \sqrt[n]{|a|^m} \left[\cos \frac{m(\theta + 2\pi k)}{n} + i \sin \frac{m(\theta + 2\pi k)}{n} \right] \quad k = 0, \ldots, n - 1$$

and so clearly, $(a^{1/n})^m \subset (a^m)^{1/n}$. Finally observe that $(a^{1/n})^m$ has n elements if and only if m and n are relatively prime, and so $(a^m)^{1/n} = (a^{1/n})^m$ if and only if m and n are relatively prime. So if m and n are relatively prime,

$$(a^m)^{1/n} = a^{m/n} = (a^{1/n})^m = \sqrt[n]{|a|^m}\left(\cos\frac{m\theta + 2\pi k}{n} + i\sin\frac{m\theta + 2\pi k}{n}\right)$$

where $k = 0, 1, 2, \ldots, n - 1$.

Example 1.11 Find and graph the n nth roots of unity, and show that their sum is zero.

Solution. Since $1 = 1(\cos 0 + i\sin 0)$, the n nth roots of 1 are given by

$$1^{1/n} = \cos\frac{2\pi k}{n} + i\sin\frac{2\pi k}{n} \qquad k = 0, 1, 2, \ldots, n - 1$$

Setting
$$\omega = \cos\frac{2\pi}{n} + i\sin\frac{2\pi}{n}$$

and using De Moivre's theorem, we find that the n nth roots of 1 are given by $1, \omega, \omega^2, \ldots, \omega^{n-1}$. Since $\omega^n = 1$ and $\omega \neq 1$, we have

$$1 + \omega + \cdots + \omega^{n-1} = \frac{\omega^n - 1}{\omega - 1} = 0$$

Clearly the n nth roots of 1 are vertices of a regular n-sided polygon inscribed in the unit circle $\{z \in \mathbf{C}: |z| = 1\}$ with one vertex at the point $z = 1$. See Fig. 1.6 for the case $n = 3$.

Fig. 1.6

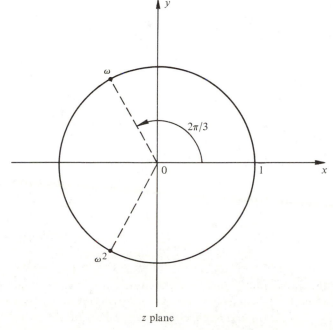

z plane

Example 1.12 Find the square roots of $-3 + 4i$.

Solution. We have

$$-3 + 4i = 5(\cos \theta + i \sin \theta)$$

where $\cos \theta = -\frac{3}{5}$ and $\sin \theta = \frac{4}{5}$. Hence

$$(-3 + 4i)^{1/2} = \sqrt{5}\left(\cos \frac{\theta + 2\pi k}{2} + i \sin \frac{\theta + 2\pi k}{2}\right) \qquad k = 0, 1$$

$$= \pm \sqrt{5}\left(\cos \frac{\theta}{2} + i \sin \frac{\theta}{2}\right)$$

As θ is an angle in the second quadrant, $\theta/2$ is an angle in the first quadrant, and

$$\cos \frac{\theta}{2} = \sqrt{\frac{1 + \cos \theta}{2}} = \frac{1}{\sqrt{5}} \qquad \sin \frac{\theta}{2} = \sqrt{\frac{1 - \cos \theta}{2}} = \frac{2}{\sqrt{5}}$$

Therefore, $(-3 + 4i)^{1/2} = \pm(1 + 2i)$.

Example 1.13 (*a*) Find $(-1)^{2/3}$ and $[(-1)^{1/3}]^2$. (*b*) Find $(-1)^{2/4}$ and $[(-1)^{1/4}]^2$.

Solution

a. $(-1)^{2/3} = [(-1)^2]^{1/3} = 1^{1/3}$

$$= \left\{1, \cos \frac{2\pi}{3} + i \sin \frac{2\pi}{3}, \cos \frac{4\pi}{3} + i \sin \frac{4\pi}{3}\right\}$$

$$= \left\{1, -\frac{1}{2} + \frac{1}{2}\sqrt{3}\, i, -\frac{1}{2} - \frac{1}{2}\sqrt{3}\, i\right\}$$

$$[-1)^{1/3}]^2 = \left\{\cos \frac{\pi}{3} + i \sin \frac{\pi}{3}, 1, \cos \frac{5\pi}{3} + i \sin \frac{5\pi}{3}\right\}^2$$

$$= \left\{\cos \frac{2\pi}{3} + i \sin \frac{2\pi}{3}, 1, \cos \frac{4\pi}{3} + i \sin \frac{4\pi}{3}\right\}$$

$$= [(-1)^2]^{1/3}$$

b. $(-1)^{2/4} = [(-1)^2]^{1/4} = 1^{1/4} = \{1, i, -1, -i\}$ while $[(-1)^{1/4}]^2 = \{1, i, -1, -i\}^2 = \{1, -1\} = 1^{1/2} \subsetneqq 1^{1/4}$.

Note in the above that 2 and 3 are relatively prime, while 2 and 4 are not relatively prime.

EXERCISES

1.5.1 Find and graph the four fourth roots of i, -16, and $-2 + 2i\sqrt{3}$.

1.5.2 Solve the equation $z^{10} + 4 = 0$.

1.5.3 Show that all the roots of the equation $(z + 1)^5 + z^5 = 0$ lie on the line $x = -\frac{1}{2}$.

1.5.4 Solve the equation $z^n = \bar{z}$.

1.5.5 Given that z_0 is an nth root of a and $\omega = \cos(2\pi/n) + i\sin(2\pi/n)$, show that $z_0, z_0\omega, z_0\omega^2, \ldots, z_0\omega^{n-1}$ are the n nth roots of a, and they are the vertices of a regular polygon.

1.5.6 Show that for $n > 0$, n an integer, the roots of the equation $(1 + z)^{2n} + (1 - z)^{2n} = 0$ are given by

$$z_k = i \tan \frac{(2k + 1)}{4n} \qquad k = 0, 1, 2, \ldots, 2n - 1$$

1.5.7 Compute $(-3 + 4i)^{-3/2}$.

1.6

QUADRATIC, CUBIC, AND QUARTIC EQUATIONS

Let f be a polynomial of degree $n \geq 1$ with complex coefficients. The equation $f(z) = 0$ is called an *algebraic equation*. To *solve* the equation $f(z) = 0$ means to find all complex numbers z such that $f(z) = 0$. These numbers z are called the *roots* or the *zeros* of the equation $f(z) = 0$. Consider the algebraic equation

$$a_0 z^n + a_1 z^{n-1} + \cdots + a_{n-1} z + a_n = 0$$

where $0 \neq a_0, a_1, a_2, \ldots, a_n$ are given complex numbers and n is a positive integer. Then n is called the *order* of the equation.

A very important theorem called the *fundamental theorem of algebra* (proved by Gauss in his Ph.D. thesis in 1799) states that every polynomial of degree n has at least one complex root. It then follows that every polynomial of degree n has exactly n roots, some of which may be repeated. A short proof of the fundamental theorem of algebra will be given in Sec. 4.4.

In this section, we shall solve algebraic equations of degrees 2, 3, and 4. (Equations of degree 1 are easily solvable in any field.) Any method that reduces the solution of an algebraic equation to a sequence of rational operations (addition, subtraction, multiplication, and division) and the extraction of nth roots of known quantities is called a *solution by radicals*. The reason that we restrict ourselves to solutions of equations of degree ≤ 4 is that algebraic equations of degree ≥ 5 are, in general, not solvable by radicals. This was proved early in the 19th century by Abel and Galois (Niels Henrik Abel, 1802–1829, Norwegian, and Évariste Galois, 1811–1832, French). The solution of quadratic equations solvable in the field of real numbers was known to the Hindus and the Greeks. The cubic equation was solved by Scipione del Ferro (1465–1526, Italian) in 1515 but did not become known until the publication of the *Ars Magna* (1545) of Geronimo Cardano (1501–1576, Italian) [who learned the proof from Niccolò Tartaglia (1500–1557, Italian) with the promise to keep it secret]. Cardano also published in *Ars Magna* the solution of the quartic equation which was discovered by his pupil Ludovico Ferrari (1522–1565, Italian).

Quadratic Equations

Consider the quadratic equation

$$az^2 + bz + c = 0 \qquad a \neq 0$$

Multiplying both sides by $4a$, we get

$$4a^2z^2 + 4abz = -4ac$$

NIELS HENRIK ABEL

Niels Henrik Abel was born in Findo, Norway, on August 5, 1802. His father was a country minister of considerable culture; his mother's outstanding characteristic seems to have been her beauty. As a student at the University of Christiania, Abel thought for a short time that he had solved by radicals the equation of degree 5, but in a famous paper published in 1824 he not only corrected his mistake but also proved the impossibility of such a solution.

Burdened with the support of his mother and five brothers and sisters when he was only 18, Abel struggled to take care of them and pursue his mathematical studies. He was unfailingly patient and cheerful, despite perennial poverty.

Abel's significant memoir on transcendental functions was presented to the Paris Academy of Sciences in 1826, but he was thoroughly ignored by all the great mathematicians of the day. Abel also contributed significantly to the theory of elliptic functions.

When he realized that he was dying of tuberculosis of the lungs, he praised the good qualities of his fiance Crelly to his friend Kielhan, and indeed Kielhan did marry Crelly after Abel's death.

Abel died on April 6, 1829, at the age of 26. Two days later a letter brought news of his appointment as professor of mathematics at the University of Berlin.

Completing the square by adding b^2 to both sides, we have

$$(2az + b)^2 = b^2 - 4ac$$

Taking square roots, we get

$$2az + b = \sqrt{b^2 - 4ac}$$

(The $\pm$ sign in front of the square root is unnecessary, since the square root of the complex number $b^2 - 4ac$ has two values.) Hence

$$z = \frac{-b + \sqrt{b^2 - 4ac}}{2a}$$

Example 1.14 Solve the equation

$$4z^2 + 4(1 + i)z + (3 - 2i) = 0$$

Solution. Here $b^2 - 4ac = 16(-3 + 4i)$, and from Example 1.12 we obtain

$$\sqrt{b^2 - 4ac} = \pm 4(1 + 2i)$$

ÉVARISTE GALOIS

Évariste Galois was born on October 25, 1811, in the small village of Burg-la-Reine near Paris, where his father served as mayor. Throughout his school years, Galois was hampered by teachers who discouraged his interest in mathematics. However, his genius refused to be quelled.

When he was 16, Galois, like Abel, believed for a short time that he had solved by radicals the equation of degree 5. Twice he took the entrance examination for the École Polytechnique, but due to the incompetence and injustice of the examiners and to his own unsystematic preparation, he failed both times. The regulations did not allow him to take the examination a third time.

Galois made fundamental discoveries in group theory. The main object of his work was to determine when a polynomial equation is solvable by radicals. However, the two important papers that he submitted for publication were lost. Most of his work was published posthumously, thanks to Joseph Liouville, who had a high regard for Galois' achievements.

Galois died in a duel on May 31, 1832, at the age of 20. Fortunately for mathematics, he spent the night before the duel composing to his friend Auguste Chevalier a long letter in which he described his mathematical discoveries.

GERONIMO CARDANO

Geronimo Cardano was born in Pavia, Italy, on September 24, 1501. He studied medicine at the University of Padua, where he received his degree in 1524. So great was Cardano's fame as a physician that in 1552 he was brought to Scotland to diagnose an ailment of the archbishop of St. Andrews. Cardano cured the archbishop of asthma by forbidding him to use feathers in his bed.

Cardano was a many-sided man. In addition to being a physician, he was an astrologer and the most notable mathematician of his day. (He was also a knave and a rascal.) He was professor of mathematics at Milan and of medicine at Pavia until 1570, when he was suddenly thrown into prison for heresy (or debt, or both). His release came within a year, and after that he lived in retirement in Rome. He published many books during his life, including works on arithmetic, games of chance, and an autobiography.

In 1545, he published the most important algebraic treatise of the Renaissance, his *Ars Magna* (the "great art," to distinguish it from the lesser art of arithmetic). In *Ars Magna*, Cardano included the solution of both cubic and quartic equations. However, he had not discovered these solutions himself. The solution of the cubic equation was discovered by Niccolò Tartaglia and entrusted to Cardano only after the latter had sworn to keep Tartaglia's method secret. In *Ars Magna* Cardano nevertheless published the method with the mere acknowledgment that a special case of this method was communicated to him by Tartaglia. Tartaglia was thus cheated of the recognition he expected to receive from publishing the solution in his own name. However, we should mention that the cubic equation was solved long before by Scipione del Ferro, who never published his results but disclosed them to his student Antonio Maria Floridus. Also, it is worth mentioning that Tartaglia had at least twice been guilty of plagiarism himself.

The solution of the quartic equation was achieved by Ludovico Ferrari, a servant and protegé of Cardano. In *Ars Magna* Cardano mentions that Ferrari solved the quartic equation at his request.

Cardano died in Rome on September 21, 1576, at the age of 74.

Hence

$$z = \frac{-4(1 + i) \pm 4(1 + 2i)}{8}$$

and the two roots are $z_1 = i/2$, $z_2 = -1 - \frac{3}{2}i$.

Cubic Equations

Consider the cubic equation

$$az^3 + bz^3 + cz + d = 0 \qquad a \neq 0 \tag{1.25}$$

The transformation $z = w - b/(3a)$ reduces (1.25) to an equation in w of the form

$$w^3 + pw + q = 0 \tag{1.26}$$

Clearly, it suffices to solve (1.26) with respect to w. Introduce two unknowns u and v with the property that

$$u + v = w \qquad \text{and} \qquad uv = -\frac{p}{3} \tag{1.27}$$

Then $u^3 + v^3 = (u + v)^3 - 3uv(u + v) = w^3 + pw = -q$ and

$$u^3 v^3 = -\frac{p^3}{27}$$

It follows that u^3 and v^3 are roots of the quadratic equation

$$t^2 + qt - \frac{p^3}{27} = 0 \tag{1.28}$$

Let t_1 and t_2 denote the two roots of (1.28). If $\sqrt{q^2/4 + p^3/27}$ denotes one of the square roots of the complex number $q^2/4 + p^3/27$, then

$$t_1 = -\frac{q}{2} + \sqrt{\frac{q^2}{4} + \frac{p^3}{27}}$$

and

$$t_2 = -\frac{q}{2} - \sqrt{\frac{q^2}{4} + \frac{p^3}{27}}$$

Now without loss of generality we have

$$u^3 = t_1 \qquad \text{and} \qquad v^3 = t_2$$

If $\sqrt[3]{t_1}$ denotes one of the three cubic roots of t_1, then the three possible values of u are (see Exercise 1.5.5)

$$u = \sqrt[3]{t_1} \qquad u = \omega\sqrt[3]{t_1} \qquad u = \omega^2\sqrt[3]{t_1}$$

where $\omega = (-1 + i\sqrt{3})/2$. Recall from Eq. (1.27) that $uv = -p/3$, and denote by $\sqrt[3]{t_2}$ that cubic root of t_2 which satisfies the relation

$$\sqrt[3]{t_1} \cdot \sqrt[3]{t_2} = -\frac{p}{3}$$

It is easy to verify that the three roots of Eq. (1.26) are given by

$$w_1 = \sqrt[3]{t_1} + \sqrt[3]{t_2} \qquad w_2 = \omega \sqrt[3]{t_1} + \omega^2 \sqrt[3]{t_2} \qquad w_3 = \omega^2 \sqrt[3]{t_1} + \omega \sqrt[3]{t_2}$$

These expressions for the roots of a cubic equation are known as the *Cardan formulas*.

Example 1.15 Solve the equation

$$z^3 + 3iz - (1 + i) = 0$$

Solution. Here

$$\frac{q^2}{4} + \frac{p^3}{27} = -\frac{i}{2}$$

One of the square roots of $-i/2$ is $\frac{1}{2}(1 - i)$. Therefore, $t_1 = 1$ and $t_2 = i$. The cube roots of t_1 are 1, ω, and ω^2, and the cube roots of t_2 are $-i$, $-i\omega$, and $-i\omega^2$, where

$$\omega = \frac{-1 + i\sqrt{3}}{2}$$

Since $1(-i) = -p/3$, we obtain from Cardan's formulas

$$z_1 = 1 - i$$

$$z_2 = \omega - i\omega^2 = \frac{(-1 - \sqrt{3}) + i(1 + \sqrt{3})}{2}$$

$$z_3 = \omega^2 - i\omega = \frac{(-1 + \sqrt{3}) + i(1 - \sqrt{3})}{2}$$

Quartic Equations
Consider the quartic equation

$$z^4 + az^3 + bz^2 + cz + d = 0 \tag{1.29}$$

Transporting the last three terms to the right-hand side and adding to both sides $(az/2)^2$, we obtain

$$\left(z^2 + \frac{a}{2}z\right)^2 = \left(\frac{a^2}{4} - b\right)z^2 - cz - d \tag{1.30}$$

If the right-hand side of Eq. (1.30) is a perfect square, we can extract square roots of both sides of (1.30), and the solution of (1.29) reduces to solving two quadratic equations. If not, adding to both sides of (1.30) the term

$$w^2 + 2\left(z^2 + \frac{a}{2}z\right)w$$

we obtain

$$\left(z^2 + \frac{a}{2}z + w\right)^2 = \left(\frac{a^2}{4} - b + 2w\right)z^2 + (aw - c)z + (w^2 - d) \quad (1.31)$$

To solve (1.31), it suffices to choose w such that the right-hand side of (1.31) becomes a perfect square. Clearly, this is the case when

$$(aw - c)^2 - 4\left(\frac{a^2}{4} - b + 2w\right)(w^2 - d) = 0 \quad (1.32)$$

Equation (1.32) is a cubic equation in w and is called the *resolvent* of the quartic equation (1.29). Choose any root of (1.32) and substitute it in (1.31). The right-hand side of (1.31) is now a perfect square. Extract square roots of both sides of (1.31) to obtain two quadratic equations that give the four roots of (1.29).

EXERCISES

1.6.1 Solve the following equations (in radicals).
    a. $4z^2 + 4(2 - i)z + (9 - 12i) = 0$ *b.* $z^3 - iz^2 + z - i = 0$
    c. $z^4 + 4iz^3 - 1 = 0$

1.6.2 In the *three-body problem* (astronomy) the following equation occurs:

$$z^3 + pz + 2 = 0 \quad \text{where } p \in \mathbf{R}$$

Show that this equation has three real roots if and only if $p \le -3$.

1.6.3 In the study of *parabolic orbits* the following equation occurs:

$$\tan \theta + {}^1\!/_3 \tan^3 \theta = t \quad (\theta \text{ and } t \text{ are real numbers})$$

Show that this equation has only one real root.

1.6.4 We say that an algebraic equation $f(z) = 0$ is of *stable type* or a *Hurwitz polynomial* if all its roots have negative real part. Show the following.
    a. The real quadratic equation (real coefficients) $z^2 + pz + q = 0$ is of stable type if and only if $p > 0$ and $q > 0$.
    b. The real cubic equation $z^3 + pz^2 + qz + r = 0$ is of stable type if and only if $p > 0$, $q > 0$, $r > 0$, and $pq > r$.
    c. The real quartic equation $z^4 + pz^3 + qz^2 + rz + s = 0$ is of stable type if and only if $p > 0$, $q > 0$, $r > 0$, $s > 0$, and $pqr > p^2s + r^2$.

Chapter 2   The Topology of
Complex Numbers

2.1
INTRODUCTION

With the geometric interpretation of the absolute value of the difference of two complex numbers in mind, we give an analytic definition of an open neighborhood of a complex number. Using this, we define various other types of subsets of $\mathbf{C}$ which turn out to have interesting properties. The definitions we give here, which extend familiar definitions dealing with the real line $\mathbf{R}$, are special cases of definitions dealing with more general spaces, namely, metric spaces. Using the completeness of $\mathbf{R}$, we prove $\mathbf{C}$ is also complete. We prove that the Cantor intersection theorem, the Bolzano-Weierstrass theorem, and the Heine-Borel theorem, familiar in $\mathbf{R}$, are also true in the complex plane $\mathbf{C}$. The concepts of convergence of sequences and series of complex numbers, as well as limits and continuity of complex functions, are also discussed here. In the final section of this chapter, the extended complex plane is introduced and some of its properties are developed.

2.2
OPEN AND CLOSED SETS

In this section, we give a few fundamental definitions. If $\varepsilon > 0$ and $z_0 \in \mathbf{C}$, the ε *neighborhood of* z_0 is denoted by $N_\varepsilon(z_0)$ and defined by

$$N_\varepsilon(z_0) = \{z \in \mathbf{C} : |z - z_0| < \varepsilon\}$$

Thus an ε neighborhood of z_0 consists of all points of the disk of radius ε centered at z_0 but excluding the boundary circle. See Fig. 2.1.

A point z_1 of a set S is called an *interior point* of S if there is an ε neighborhood of z_1 which is contained in the set S (see Fig. 2.1). The set of all interior points of S, denoted by $\overset{o}{S}$, is called the *interior* of S.

A point z_0 is called an *exterior point* of S if there is an ε neighborhood of z_0 that contains no points of S (see Fig. 2.1). The set of all exterior points of S is called the *exterior* of S.

A point z_2 which is neither an interior nor an exterior point of S is called a *boundary point* of S (see Fig. 2.1). The set of all boundary points of S is denoted by $bd\ S$ and is called the *boundary* of S. Clearly,

$$\overset{o}{N_\varepsilon}(z_0) = N_\varepsilon(z_0)$$

the exterior of $N_\varepsilon(z_0) = \{z \in \mathbf{C} : |z - z_0| > \varepsilon\}$, and

$$bd\ N_\varepsilon(z_0) = \{z \in \mathbf{C} : |z - z_0| = \varepsilon\}$$

A set S is called *open* if it consists entirely of interior points, that is, $S = \overset{o}{S}$. A set S is called *closed* if its complement $\mathbf{C} - S$ is open. Clearly, the empty set $\varnothing$ and the whole complex plane $\mathbf{C}$ are both open and closed sets. Also the sets $\{z \in \mathbf{C} : |z - i| < 1\}$ and $\{z \in \mathbf{C} : |z - 2 + i| > 3\}$ are open, whereas the sets $\{z \in \mathbf{C} : |z| \leq 2\}$ and $\{z \in \mathbf{C} : |z - 1| \geq 1\}$ are closed. However, the set $\{z \in \mathbf{C} : 1 \leq |z - i| < 3\}$ is neither closed nor open.

Fig. 2.1

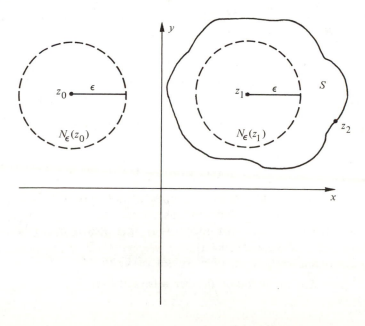

A set S is called *bounded* if there exists a constant M such that $|z| < M$ for all $z \in S$. That is, $S \subset N_M(0)$.

A subset of $\mathbf{C}$ that is both closed and bounded is called *compact*. For example, the set $\{z \in \mathbf{C} : |z - 2i| \leq 3\}$ is compact, but the sets $\{z \in \mathbf{C} : |z - 2i| < 3\}$ and $\{z \in \mathbf{C} : |z - 2i| \geq 3\}$ are not compact.

A point z is called a *limit point* of a set S if each ε neighborhood of z contains at least one point of S different from z. (A limit point of S need not belong to S.) The set of limit points of S, denoted by S', is called the *derived set* of S. The union $S \cup S'$, denoted by $\bar{S}$, is called the *closure* of S. A point $z_0 \in S$ is called an *isolated point* of S if there is an ε neighborhood of z_0 containing no other points of S, i.e., if the intersection $S \cap N_\varepsilon(z_0)$ consists of the singleton $\{z_0\}$.

As an example, let $S = \{(-1)^n + (ni)/(n + 1) : n \in \mathbf{N}\}$ where $\mathbf{N}$ denotes the set of positive integers. The points $1 + i$ and $-1 + i$ are the only limit points of S, and so $S' = \{1 + i, 1 - i\}$. Clearly every point of S is an isolated point. Here $\bar{S} = S \cup \{1 + i, 1 - i\}$.

The *diameter* of a nonempty set S is denoted by $d(S)$ and defined by

$$d(S) = \sup\{|z_1 - z_2| : z_1, z_2 \in S\}$$

Clearly the diameter of an ε neighborhood is 2ε, and the diameter of a singleton is 0. Also, a nonempty set S is bounded if and only if $d(S) < \infty$.

The *distance* between two nonempty sets S_1 and S_2 is denoted by $d(S_1, S_2)$ and defined by

$$d(S_1, S_2) = \inf\{|z_1 - z_2| : z_1 \in S_1 \text{ and } z_2 \in S_2\}$$

The *distance* from a point z to a nonempty set S, denoted by $d(z, S)$, is defined by $d(z, S) = d(\{z\}, S)$. As an example, let $S_1 = \{z \in \mathbf{C} : \operatorname{Re} z \geq 3\}$ and $S_2 = \{z \in \mathbf{C} : |z - i| < 2\}$. Then $d(S_1) = \infty$, $d(S_2) = 4$, $d(S_1, S_2) = 1$, $d(3i, S_2) = 0$, and $d(i, S_1) = 3$.

Remark 2.1 In the future for convenience we shall often omit the braces in defining sets in $\mathbf{C}$ and write only the property characterizing the set. For example, we shall say the set $|z - z_0| < \varepsilon$ is an ε neighborhood of z_0, and its boundary the set $|z - z_0| = \varepsilon$, etc.

EXERCISES

2.2.1 Show that a subset G of $\mathbf{C}$ is open if and only if G is a (possibly infinite) union of ε neighborhoods.

2.2.2 Show that any union of open sets is open and that any finite intersection of open sets is open.

2.2.3 Show that if G is an open subset of $\mathbf{C}$ and $G \subset S$, then $G \subset \overset{o}{S}$. Hence $\overset{o}{S}$ is the union of all open sets contained in S.

2.2.4 Show that any intersection of closed sets is closed and that any finite union of closed sets is closed.

2.2.5 Show that if F is a closed subset of C and $S \subset F$, then $\bar{S} \subset F$. Hence $\bar{S}$ is the intersection of all closed sets containing S.

2.2.6 Show that a set S is closed if and only if $S' \subset S$.

2.2.7 Show that a set S is closed if and only if $S = \bar{S}$.

2.2.8 Show that for any subset S of C, the sets bd S, S', and $\bar{S}$ are closed.

2.2.9 Show that any finite union of compact sets is compact and that any intersection of compact sets is compact.

2.2.10 Show that any set S consisting of a finite number of points is compact.

2.3

FUNCTIONS AND CONTINUITY

Let A and B be sets. A subset f of the Cartesian product $A \times B$ is called a *function* (or *mapping*) from A into B, and it is written $f: A \to B$, if for each $z \in A$ there exists exactly one $w \in B$ such that $(z,w) \in f$. When $(z,w) \in f$, we agree to write $w = f(z)$, and we call w the *image* of z under f, or the *value* of f at z. We call A the *domain* of f, and we write dmn $f = A$; similarly, we call $\{f(z): z \in A\}$ the *image* of f and write im $f = \{f(z): z \in A\}$.

Let A_1 and B_1 be sets and $f: A \to B$. We define $f[A_1] = \{f(z): z \in A_1 \cap A\}$ and $f^{-1}[B_1] = \{z \in A: f(z) \in B_1\}$. The set $f[A_1]$ is called the *image* of A_1 under f, and the set $f^{-1}[B_1]$ is called the *inverse image* of B_1 under f.

Let $f: A \to B$ be a function. If $A \subset C$, f is called a function of a *complex variable* and if $B \subset C$, f is called a *complex-valued* function.

The above definition of a complex-valued function has traditionally been called a *single-valued function*, in contrast to functions like the argument function or the nth-root function, with which the reader is familiar from Chapter 1, and which have traditionally been called *multiple-valued*.

The function $f: A \to B$ is called *onto* if $f[A] = B$; it is called *one-to-one* if for every $z_1, z_2 \in A$ with $z_1 \ne z_2$, $f(z_1) \ne f(z_2)$. Note that if $f: A \to B$ is one-to-one and onto, then $f^{-1} = \{(b,a): (a,b) \in f\}$ is a function from B onto A, and in fact $f^{-1}: B \to A$ is one-to-one and onto. The function f^{-1} is called the *inverse* of f.

Let $f: A \to C$ and $g: B \to C$ be functions. Set $D = A \cap f^{-1}[B]$. The *composition* of g with f is the function $g \circ f: D \to C$ defined by $(g \circ f)(a) = g(f(a))$. In particular, if $f: A \to B$ and $g: B \to C$, then $g \circ f: A \to C$ is given by $(g \circ f)(a) = g(f(a))$. Note that in general, $f \circ g \ne g \circ f$.

Let $f: A \to B$ and let $z_0 \in A'$. We say that f has a *limit* l as z approaches z_0, and we write

$$\lim_{z \to z_0} f(z) = l$$

if for every ε neighborhood $N_\varepsilon(l)$ of l there exists a δ neighborhood $N_\delta(z_0)$ of z_0 such that for all $z \in A \cap N_\delta(z_0)$ with $z \ne z_0$, $f(z) \in N_\varepsilon(l)$. This definition is equivalent to the following: For every $\varepsilon > 0$, there exists $\delta > 0$ such that if $z \in A$ and $0 < |z - z_0| < \delta$, then $|f(z) - l| < \varepsilon$.

As in the case of real functions, one can show that if a limit exists, it is unique. Also if $f: A \to \mathbf{C}$ and $F: B \to \mathbf{C}$ with $z_0 \in A' \cap B'$ such that

$$\lim_{z \to z_0} f(z) = l \quad \text{and} \quad \lim_{z \to z_0} F(z) = L$$

then

$$\lim_{z \to z_0} [af(z) + bF(z)] = al + bL \quad \text{for every } a, b \in \mathbf{C} \quad (2.1)$$

$$\lim_{z \to z_0} [f(z)F(z)] = lL \quad (2.2)$$

$$\lim_{z \to z_0} \frac{f(z)}{F(z)} = \frac{l}{L} \quad L \neq 0 \quad (2.3)$$

Example 2.1 Let $f: \mathbf{C} - \{3i\} \to \mathbf{C}$ be defined by

$$f(z) = \frac{z^2 + 9}{4(z - 3i)}$$

Show that $\lim_{z \to 3i} f(z) = 3i/2$.

Proof. Observe that $3i$ is a limit point of the domain of f. If z is in the domain of f, we have

$$|f(z) - 3i| = \left| \frac{z^2 + 9}{4(z - 3i)} - \frac{3i}{2} \right| = \frac{1}{4} |z - 3i|$$

Let $\varepsilon > 0$ be given. We must choose $\delta > 0$ such that $0 < |z - 3i| < \delta$ implies $\frac{1}{4}|z - 3i| < \varepsilon$. Clearly δ may be chosen to be any positive number less than or equal to 4ε. For example, let us choose $\delta = 4\varepsilon$.

Let $f: A \to B$ and let $z_0 \in A$. We say that f is *continuous* at z_0 if either $z_0 \notin A'$ or $z_0 \in A'$ and $\lim_{z \to z_0} f(z)$ exists and is equal to $f(z_0)$. If f is continuous at every point of A, then f is called *continuous*. f is called *discontinuous* at z_0 if it is not continuous at z_0. If f is discontinuous at z_0 but $\lim_{z \to z_0} f(z)$ exists, we can "make" f continuous at z_0 by redefining $f(z_0) = \lim_{z \to z_0} f(z)$. Such a discontinuity of f is called *removable*. Note that if f is continuous at z_0 and G is an open set containing $f(z_0)$, then there exists $\delta > 0$ such that

$$A \cap N_\delta(z_0) \subset f^{-1}(G)$$

It follows that $f: A \to B$ is continuous if and only if for every open set G_1 in $\mathbf{C}$ there exists an open set G_2 in $\mathbf{C}$ such that $f^{-1}[G_1] = A \cap G_2$.

As in the case of real functions (or by using the previous properties of limits), one can show that if $f: A \to \mathbf{C}$ and $F: B \to \mathbf{C}$ are continuous at $z_0 \in A \cap B$, and if $a, b \in \mathbf{C}$, then the function $af + bF$ is continuous at z_0. If $F(z_0) \neq 0$, then f/F is also continuous at z_0. If f is continuous at z_0,

$f(z_0)$ is an element of B, and F is continuous at $f(z_0)$, then the composite function $F \circ f$ is again continuous at z_0.

Example 2.2 Let $f: \mathbf{C} \to \mathbf{C}$ be given by

$$f(z) = \begin{cases} \dfrac{z^2 + 9}{4(z - 3i)} & z \neq 3i \\[4mm] \dfrac{3i}{2} & z = 3i \end{cases}$$

Show that f is continuous on $\mathbf{C}$.

Proof. The continuity of f at $z_0 = 3i$ follows from Example 2.1, since $\lim_{z \to 3i} f(z)$ exists and is equal to $3i/2 = f(3i)$. The proof of the continuity of f at any other point of $\mathbf{C}$ is left to the student. (It suffices to show that both the numerator and the denominator of f are continuous.)

A function $f: A \to B$ is called *uniformly continuous* on A if, for every $\varepsilon > 0$, there exists $\delta > 0$ such that for all $z_1, z_2 \in A$ with $|z_1 - z_2| < \delta$, we have $|f(z_1) - f(z_2)| < \varepsilon$. Clearly if f is uniformly continuous on A, it is also continuous, but the converse is not true.

Example 2.3 Let D be the punctured disk $0 < |z| < 1$, and let $f: D \to \mathbf{C}$ be defined by $f(z) = \bar{z}^2$. Show that f is uniformly continuous on D.

Proof. In fact, if $z_1, z_2 \in D$, we have

$$\begin{aligned} |f(z_1) - f(z_2)| = |\bar{z}_1^2 - \bar{z}_2^2| &= |\bar{z}_1 + \bar{z}_2| \, |\bar{z}_1 - \bar{z}_2| \\ &\leq (|z_1| + |z_2|)|z_1 - z_2| \leq 2|z_1 - z_2| \end{aligned}$$

Let $\varepsilon > 0$ be given. We must choose $\delta > 0$ so that if $z_1, z_2 \in D$ and $|z_1 - z_2| < \delta$, then $|f(z_1) - f(z_2)| < \varepsilon$. Clearly $\delta = \varepsilon/2$ is a good choice, as $|f(z_1) - f(z_2)| < 2\varepsilon/2 = \varepsilon$.

Remark 2.2 Let $f: A \to \mathbf{C}$ and $g: B \to \mathbf{C}$. In the future, unless otherwise specified, the notation $w = f(z)/g(z)$ or even $f(z)/g(z)$ stands for the function with domain $D = A \cap B \cap \{z: g(z) \neq 0\}$ whose value at $z \in D$ is $f(z)/g(z)$.

By separating it into real and imaginary parts, we can express any function f in terms of two real functions u and v by setting $f(x + iy) = u(x,y) + iv(x,y)$. For example, the function $w = z^2 + 3iz$ is equivalent to

$$u = x^2 - y^2 - 3y \quad \text{and} \quad v = 2xy + 3x$$

This is especially useful in plotting a complex-valued function of a complex variable, where a whole plane is needed to represent the domain of the function and another plane to represent the image of the function. For example, to find the image of a set A under the mapping $w = z^2$, we consider A as a subset of a plane called "z plane" and its image as a subset of another plane called "w plane."

Example 2.4 Find the image of the strip $1 \leq \mathrm{Re}\, z \leq 2$ under the mapping $w = z^2$.

Solution. Set $z = x + iy$ and $w = u + iv$. Then the mapping $w = z^2$ is equivalent to

$$u = x^2 - y^2 \qquad v = 2xy$$

The line $x = x_0$ (see Fig. 2.2) is mapped onto the curve

$$u = x_0^2 - y^2 \qquad v = 2x_0 y \qquad -\infty < y < \infty$$

For $x_0 \neq 0$, eliminating y gives us

$$u = x_0^2 - \frac{v^2}{4x_0^2} \qquad -\infty < v < +\infty$$

which is a parabola in the w plane. As x_0 varies from 1 to 2, the lines $x = x_0$ sweep out the strip $1 \leq \mathrm{Re}\, z \leq 2$, and the above parabolas sweep out the shaded region shown in Fig. 2.2. The boundary of the image consists of the parabolas

$$u = 1 - \frac{v^2}{4} \qquad \text{and} \qquad u = 4 - \frac{v^2}{16}$$

Fig. 2.2

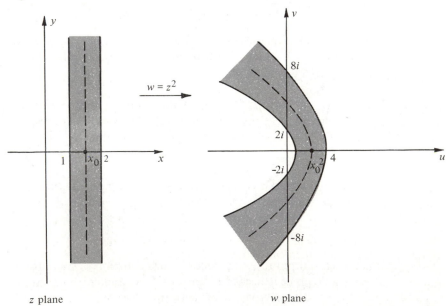

z plane

w plane

EXERCISES

2.3.1 Compute the limit (if it exists).

    a. $\lim\limits_{z \to 0} \dfrac{\bar{z}}{z}$ *b.* $\lim\limits_{z \to -1} \dfrac{z^4 - 2z^2 + 1}{z + 1}$

2.3.2 Find where the following functions are continuous and which discontinuities are removable.

    a. $(\operatorname{Re} z)/z$ *b.* $\operatorname{Im}[z/(z - 1)]$

    c. $\operatorname{Re} z^2$ *d.* $(z \operatorname{Re} z)/|z|$

2.3.3 Show that the function $f(z) = 1/z$ is uniformly continuous in $\frac{1}{3} \le |z| < 1$ but not in $0 < |z| \le 1$.

2.3.4 Find the images of the strip $1 < |\operatorname{Im} z| \le 2$ and the disk $|z| < 1$ under the mappings in (*a*) and (*b*).

    a. $w = z^2$ *b.* $w = \dfrac{2z + i}{z + 1}$

2.3.5 Let $f(z) = u(x,y) + iv(x,y)$. Also let $l = l_1 + il_2$ and $z_0 = x_0 + iy_0$. Show that

    a. $\lim\limits_{z \to z_0} f(z) = l$ if and only if $\lim\limits_{(x,y) \to (x_0, y_0)} u(x,y) = l_1$ and

         $\lim\limits_{(x,y) \to (x_0, y_0)} v(x,y) = l_2$

    b. f is continuous at z_0 if and only if u and v are both continuous at (x_0, y_0)

2.4

SEQUENCES AND SERIES OF
COMPLEX NUMBERS

Let $\mathbf{N}$ denote the set of positive integers. A *sequence* of complex numbers is a function $z: \mathbf{N} \to \mathbf{C}$. If z is a sequence of complex numbers, it is customary to denote the ordered pair $(n, z(n))$ by z_n and the sequence z by $\{z_n\}$. Other common notations for the sequence z are $\{z_n\}_{n=1}^{\infty}$ or $z_1, z_2, \ldots$.

Let $z_1, z_2, \ldots$ be a sequence of complex numbers. If $\{n_1, n_2, \ldots\}$ is a subset of $\mathbf{N}$ such that $n_1 < n_2 < \cdots$, then the sequence $z_{n_1}, z_{n_2}, \ldots$ is called a *subsequence* of $\{z_n\}$.

Let $\{z_n\}$ be a sequence of complex numbers. We say that $\{z_n\}$ *converges* to a point $z \in \mathbf{C}$ and we write $\lim_{n \to \infty} z_n = z$ if, for each $\varepsilon > 0$, there exists an integer n_0 such that $n \ge n_0$ implies $|z_n - z| < \varepsilon$. As in the case of real sequences, a sequence $\{z_n\}$ of complex numbers can converge to at most one point z. The point z is then called the *limit* of the sequence. The convergence of a sequence can be also defined in terms of ε neighborhoods, as follows: $\lim_{n \to \infty} z_n = z$ if, for every ε neighborhood $N_\varepsilon(z)$ of z, there is an integer n_0 such that $n \ge n_0$ implies $z_n \in N_\varepsilon(z)$.

If a sequence fails to converge, it is said to *diverge*.

A sequence $\{z_n\}$ is called a *Cauchy sequence* if, for each $\varepsilon > 0$, there exists an integer n_0 such that $n, m \geq n_0$ implies $|z_n - z_m| < \varepsilon$. In Exercise 1.4.11, we proved that the complex plane $\mathbf{C}$, with the distance between any two points z_1, z_2 defined by $|z_1 - z_2|$, is a metric space. A metric space with the additional property that every Cauchy sequence converges (and the limit belongs to the space) is called a *complete metric space*. For example, the real line $\mathbf{R}$ with the usual distance is a complete metric space. The completeness of $\mathbf{C}$ is a direct consequence of the completeness of $\mathbf{R}$. In fact, we can prove the following theorem.

Theorem 2.1 Let $\{z_n\}$ be a sequence of complex numbers. Then the following statements are equivalent:

$$\{z_n\} \text{ converges to some complex number } z \tag{2.4}$$

$$\{z_n\} \text{ is a Cauchy sequence} \tag{2.5}$$

Proof. Statement (2.4) implies (2.5). This follows from the triangle inequality

$$|z_n - z_m| \leq |z_n - z| + |z - z_m|$$

Also, (2.5) implies (2.4). Let us set

$$z_n = x_n + iy_n \qquad n = 1, 2, \ldots$$

Clearly,

$$|x_n - x_m| \leq |z_n - z_m|$$

and

$$|y_n - y_m| \leq |z_n - z_m|$$

From these inequalities it follows that the real sequences $\{x_n\}$ and $\{y_n\}$ are Cauchy sequences in $\mathbf{R}$. By the completeness of $\mathbf{R}$ there exist real numbers x and y such that

$$\lim_{n \to \infty} x_n = x \qquad \text{and} \qquad \lim_{n \to \infty} y_n = y$$

Set $z = x + iy$ and observe that

$$|z_n - z| = |(x_n - x) + i(y_n - y)| \leq |x_n - x| + |y_n - y|$$

Hence

$$\lim_{n \to \infty} z_n = z$$

and the proof is complete.

Theorem 2.2 Let

$$z = x + iy \qquad \text{and} \qquad z_n = x_n + iy_n \quad \text{for } n = 1, 2, \ldots$$

Then the following statements are equivalent:

$$\lim_{n \to \infty} z_n = z \tag{2.6}$$

$$\lim_{n \to \infty} x_n = x \qquad \text{and} \qquad \lim_{n \to \infty} y_n = y \tag{2.7}$$

Proof. Statement (2.6) implies (2.7). This follows from the inequalities

$$|x_n - x| \le |z_n - z| \qquad \text{and} \qquad |y_n - y| \le |z_n - z|$$

Statement (2.7) implies (2.6). This follows from the inequality

$$|z_n - z| \le |x_n - x| + |y_n - y|$$

The proof is complete.

As in the case of real sequences, it is easy to establish that if

$$\lim_{n \to \infty} z_n = z \qquad \text{and} \qquad \lim_{n \to \infty} \zeta_n = \zeta$$

then

$$\lim_{n \to \infty} (az_n + b\zeta_n) = az + b\zeta \qquad \text{for every } a, b \in \mathbf{C} \tag{2.8}$$

$$\lim_{n \to \infty} (z_n \zeta_n) = z\zeta \tag{2.9}$$

$$\lim_{n \to \infty} \frac{z_n}{\zeta_n} = \frac{z}{\zeta} \qquad \zeta \ne 0 \tag{2.10}$$

Let $\{z_n\}$ be a sequence of complex numbers. The symbol

$$z_1 + z_2 + \cdots + z_n + \cdots$$

is called an *infinite series* of complex numbers, and is denoted by

$$\sum_{n=1}^{\infty} z_n$$

The complex number

$$S_n = z_1 + z_2 + \cdots + z_n$$

is called the nth partial sum of the series $\sum_{n=1}^{\infty} z_n$.

We say that the series $\sum_{n=1}^{\infty} z_n$ *converges* if there exists a complex number S such that

$$\lim_{n \to \infty} S_n = S$$

S is then called the *sum* of the series, and we write

$$\sum_{n=1}^{\infty} z_n = S$$

If $\lim_{n \to \infty} S_n$ does not exist, then the series is said to *diverge*. The following result is a consequence of Theorem 2.2.

Corollary 2.1 Let

$$z_n = x_n + iy_n \quad n = 1, 2, \ldots \qquad \text{and} \qquad S = U + iV$$

Then the following statements are equivalent:

$$\sum_{n=1}^{\infty} z_n = S \tag{2.11}$$

$$\sum_{n=1}^{\infty} x_n = U \qquad \text{and} \qquad \sum_{n=1}^{\infty} y_n = V \tag{2.12}$$

As in the case of real series, it is easy to establish that if

$$\sum_{n=1}^{\infty} z_n = S \quad \text{and} \quad \sum_{n=1}^{\infty} \zeta_n = T$$

then, for every $a, b \in \mathbf{C}$,

$$\sum_{n=1}^{\infty} (az_n + b\zeta_n) = aS + bT$$

The series $\sum_{n=1}^{\infty} z_n$ is called *absolutely convergent* if the series

$$\sum_{n=1}^{\infty} |z_n|$$

converges. A series that converges but is not absolutely convergent is called *conditionally convergent*.

The proof for the following theorem holds unchanged from the corresponding proof for real series.

Theorem 2.3 Let $\sum_{n=1}^{\infty} z_n$ be an absolutely convergent series with sum S. Then every rearrangement of $\sum_{n=1}^{\infty} z_n$ converges to S.

The next result (whose proof follows from the generalized triangle inequality) is often useful in estimating the sum of a series.

Theorem 2.4 Let $\sum_{n=1}^{\infty} z_n$ be an absolutely convergent series. Then

$$\left| \sum_{n=1}^{\infty} z_n \right| \leq \sum_{n=1}^{\infty} |z_n|$$

The tests for absolute convergence of real series (*comparison test, ratio test, root test*, etc.) apply unchanged to complex series too. Here we prove the root test; we shall leave the other tests for the exercises. Let us first remind the student of the concept of *limit superior* (*limit inferior*) of a sequence $\{a_n\}$ of real numbers, which is denoted by

$$\lim_{n \to \infty} \sup_{k \geq n} a_k \qquad \left(\lim_{n \to \infty} \inf_{k \geq n} a_k \right)$$

Other customary notations are

$$\lim_{n \to \infty} \sup a_n, \qquad \overline{\lim_{n \to \infty}} \, a_n \qquad \left(\lim_{n \to \infty} \inf a_n, \qquad \underline{\lim_{n \to \infty}} \, a_n \right)$$

Let $\{a_n\}$ be a bounded sequence of real numbers. We say that $\lim_{n \to \infty} \sup_{k \geq n} a_k = L$ if for every $\varepsilon > 0$ the following two conditions hold:

$$a_n < L + \varepsilon \qquad \text{for all but a finite number of values of } n \qquad (2.13)$$

$$a_n > L - \varepsilon \qquad \text{for infinitely many values of } n \qquad (2.14)$$

Similarly we define $\lim_{n \to \infty} \inf_{k \geq n} a_k = l$ if for every $\varepsilon > 0$ the following two conditions hold:

$$a_n > l - \varepsilon \qquad \text{for all but a finite number of values of } n \qquad (2.13a)$$

$$a_n < l + \varepsilon \qquad \text{for infinitely many values of } n \qquad (2.14a)$$

If the sequence $\{a_n\}$ is not bounded from above (from below), then we define

$$\lim_{\substack{n \to \infty \\ k \geq n}} \sup a_k = +\infty \qquad \left(\lim_{\substack{n \to \infty \\ k \geq n}} \inf a_k = -\infty \right)$$

It is not difficult to show that if $\{a_n\}$ is a bounded sequence of real numbers and S is the set of limits of all convergent subsequences of $\{a_n\}$, then

$$\lim_{\substack{n \to \infty \\ k \geq n}} \sup a_n = \sup S \qquad \text{and} \qquad \lim_{\substack{n \to \infty \\ k \geq n}} \inf a_k = \inf S$$

It is an easy check that if $a_1, a_2, \ldots$ is a sequence of real numbers, then

$$\lim_{\substack{n \to \infty \\ k \geq n}} \sup a_k = \lim_{n \to \infty} \sup \{a_k : k \geq n\}$$

and

$$\lim_{\substack{n \to \infty \\ k \geq n}} \inf a_k = \lim_{n \to \infty} \inf \{a_k : k \geq n\}$$

If $\overline{\lim}_{n \to \infty} a_n = \underline{\lim}_{n \to \infty} a_n = L$, then clearly $\lim_{n \to \infty} a_n = L$. Conversely if $\lim_{n \to \infty} a_n = L$, then $\overline{\lim}_{n \to \infty} a_n = \underline{\lim}_{n \to \infty} a_n = L$. Clearly the limit superior and limit inferior of any sequence always exist, but the limit of the sequence $\{a_n\}$ may not exist. For example, let $a_n = \frac{1}{3}[(-1)^n + \sin(n\pi/2)]$. Then we have $\lim_{n \to \infty} \sup_{k \geq n} a_k = +\frac{2}{3}$ and $\lim_{n \to \infty} \inf_{k \geq n} a_k = -\frac{2}{3}$, but $\lim_{n \to \infty} a_n$ does not exist.

Theorem 2.5. nth-root Test Let $\lim_{n \to \infty} \sup_{k \geq n} \sqrt[k]{|z_k|} = L$. Then the series $\sum_{n=1}^{\infty} z_n$ converges absolutely if $L < 1$ and diverges if $L > 1$.

Proof. Assuming $L < 1$, choose an r such that $L < r < 1$. Then $\sqrt[n]{|z_n|} \leq r$ for all sufficiently large n, say $n \geq n_0$. Thus $|z_n| \leq r^n$ if $n \geq n_0$, and so $\sum_{n=1}^{\infty} |z_n|$ converges by comparing it with the geometric series $\sum_{n=1}^{\infty} r^n$, which converges because $r < 1$. If $L > 1$, then from the above definition of limit superior, we see that $\sqrt[n]{|z_n|} \geq 1$ for infinitely many values of n. Thus $|z_n| \geq 1$ for infinitely many values of n and hence, by Exercise 2.4.4, $\sum_{n=1}^{\infty} |z_n|$ diverges.

Example 2.5 By using the nth-root test, we can easily see that the series

$$\sum_{n=1}^{\infty} \frac{[(-1)^n + \sin(n\pi/2)]^n}{3^n}$$

converges absolutely.

EXERCISES

2.4.1 Show that a point z is a limit point of a set S if and only if there exists a sequence of distinct points in S that converges to z.

2.4.2 Given that $\lim_{n \to \infty} z_n = z$, show that

$$\lim_{n \to \infty} \frac{z_1 + z_2 + \cdots + z_n}{n} = z$$

2.4.3 Evaluate the following limits if they exist.

    a. $\displaystyle \lim_{n \to \infty} \frac{n! \, i^n}{n^n}$ *b.* $\displaystyle \lim_{n \to \infty} i^n$ *c.* $\displaystyle \lim_{n \to \infty} n \left(\frac{1 + i}{2} \right)^n$

    d. $\displaystyle \lim_{n \to \infty} \sup_{k \geq n} \left[(-1)^k + \frac{k}{k+1} \right]$ *e.* $\displaystyle \lim_{k \to \infty} \inf_{n \geq k} (-1)^n n$

2.4.4 Given that $\sum_{n=1}^{\infty} z_n$ converges, show that $\lim_{n \to \infty} z_n = 0$ but that the converse is not necessarily true.

2.4.5 *Comparison test.* Let $0 \leq a_n \leq |z_n| \leq b_n$ for $n = 1, 2, \ldots$. Show the following.
    a. If $\sum_{n=1}^{\infty} a_n$ diverges, then the series $\sum_{n=1}^{\infty} z_n$ diverges.
    b. If $\sum_{n=1}^{\infty} b_n$ converges, then the series $\sum_{n=1}^{\infty} z_n$ converges absolutely.

2.4.6 *Ratio test.* Let $\lim_{n \to \infty} \sup_{k \geq n} |z_{k+1}/z_k| = L$. Show that the series $\sum_{n=1}^{\infty} z_n$ converges absolutely if $L < 1$, diverges if $L > 1$, and that no conclusion can be reached from the ratio test when $L = 1$.

2.4.7 Test the following series for convergence, absolute convergence, conditional convergence, and divergence.

    a. $\displaystyle \sum_{n=1}^{\infty} \left(\frac{n}{n+1} \right)^{n^2} i^n$ *b.* $\displaystyle \sum_{n=1}^{\infty} \frac{\log n}{n!} i^n$

    c. $\displaystyle \sum_{n=1}^{\infty} \frac{(\sqrt{3} + i)^n}{5^{n/2}}$ *d.* $\displaystyle \sum_{n=2}^{\infty} \frac{n i^n}{\log n}$

    e. $\displaystyle \sum_{n=1}^{\infty} \frac{(-1)^n}{n+i}$

2.4.8 Let $\sum_{n=1}^{\infty} z_n = S$ and $\sum_{n=1}^{\infty} \zeta_n = T$ be absolutely convergent series. Set $a_n = z_1 \zeta_n + z_2 \zeta_{n-1} + \cdots + z_n \zeta_1$ and show that the series $\sum_{n=1}^{\infty} a_n$ converges absolutely and has sum ST.

2.4.9 Show that the ratio test fails for the series $\sum_{n=1}^{\infty} z_n$, where

$$z_n = \begin{cases} \dfrac{i}{2^k} & n = 2k \\[2ex] \dfrac{i}{2^{k+1}} & n = 2k - 1 \end{cases}$$

but that the nth-root test works.

In Theorem 2.1 we proved that $\mathbf{C}$ is complete; that is, every Cauchy sequence of complex numbers converges in $\mathbf{C}$. Using this property of $\mathbf{C}$, we now prove the following theorem of Cantor (Georg Cantor, 1845–1918, Russian).

Theorem 2.6. *Cantor's Intersection Theorem* Let $K_1 \supset K_2 \supset \cdots \supset K_n \supset \cdots$ be a decreasing sequence of nonempty closed subsets of $\mathbf{C}$ such that $\lim_{n \to \infty} d(K_n) = 0$. Then the set $K = \bigcap_{n=1}^{\infty} K_n$ contains exactly one point.

Proof. It is clear that because $\lim_{n \to \infty} d(K_n) = 0$, K cannot contain more than one point. Therefore it suffices to show that K is nonempty. For each $n \in \mathbf{N}$, choose a point $z_n \in K_n$. Then because $\lim_{n \to \infty} d(K_n) = 0$, clearly $\{z_n\}$ is a Cauchy sequence. Since $\mathbf{C}$ is complete, there exists a point $z \in \mathbf{C}$ such that $\lim_{n \to \infty} z_n = z$. We shall prove that $z \in K$. For this it is enough to show that for any given $n_0 \in \mathbf{N}$, $z \in K_{n_0}$. If the sequence $\{z_n\}$ has only a finite number of distinct points, then because $\lim_{n \to \infty} z_n = z$ there exists $n_1 \in \mathbf{N}$ such that

---

GEORG CANTOR

Georg Cantor was born of Danish parents in St. Petersburg, Russia, on March 3, 1845, but his family moved to Germany in 1856. His father was a successful merchant and his mother came from a family of artists.

Cantor studied at the universities of Zurich, Berlin, and Göttingen and received his doctorate from the University of Berlin in 1867. Cantor's ambition to become a professor at the University of Berlin never materialized. He spent all his professional career at the third-rate University of Halle.

Cantor's main contribution to mathematics was the creation of set theory. His first revolutionary paper on the subject was published in 1874. However, his theory was badly attacked by the influential Leopold Kronecker as being "mathematically insane." After many years, Cantor's theory of sets was not only accepted but recognized as a fundamental contribution to all mathematics.

Cantor married Vally Guttman in 1874; six children were born of the marriage. During the last 34 years of his life, Cantor suffered several nervous breakdowns, and he died on January 6, 1918, in a mental hospital.

if $n \geq n_1$, $z_n = z$. Hence $z = z_{n_0 n_1} \in K_{n_0 n_1} \subset K_{n_0}$. If the sequence $\{z_n\}$ has infinitely many distinct points, then z is a limit point of the set of these distinct points and therefore a limit point of K_{n_0}. By Exercise 2.2.6, we saw that a closed set contains all its limit points. Hence $z \in K_{n_0}$ and the proof is complete.

A basic property of the real numbers is the *Bolzano-Weierstrass property* (Bernhard Bolzano, 1781–1848, Czechoslovakian), which states that a bounded infinite set of real numbers has at least one limit point. From this it follows that every bounded sequence of real numbers has a convergent subsequence.

The following theorem proves that the Bolzano-Weierstrass property is also true for **C**.

Theorem 2.7. *Bolzano-Weierstrass Theorem* Every bounded infinite set S of complex numbers has a limit point.

Proof. Let $\{z_n\}$ be a sequence of distinct points in S. Let $z_n = x_n + iy_n$ for $n = 1, 2, \ldots$. Since S is bounded, there exists a constant M such that $|z_n| \leq M$ for all $n \in \mathbf{N}$. Hence $|x_n| \leq M$ and $|y_n| \leq M$ for all $n \in \mathbf{N}$. The sequence $\{x_n\}$ has a subsequence $\{x_{n(k)}\}_{k=1}^{\infty}$ which converges, say to x_0. The

---

BERNHARD BOLZANO

Bernhard Bolzano was born in Prague (in what is now Czechoslovakia) on October 5, 1781. His father, an Italian emigrant, was an art dealer, and his mother was the daughter of a hardware tradesman in Prague. Bolzano studied philosophy, physics, mathematics, and theology at the University of Prague, and in 1807 he was appointed professor of the philosophy of religion at the same university. In 1816, he was accused of heresies and, as a result, was forbidden to teach or publish from 1819 to 1825.

Bolzano published many works on analysis, geometry, logic, philosophy, and religion. He used the so-called "Bolzano-Weierstrass theorem" in his *Functionenlehre*. He did not prove the theorem there but referred to his previous work (in which, until now, the theorem has not been found). Bolzano's main work is *Paradoxien des Unendlichen*. This treatise, which was published posthumously in 1850, contains many properties of infinite sets.

Bolzano died in Prague on December 18, 1848.

---

subsequence $\{y_{n(k)}\}_{k=1}^{\infty}$ of $\{y_n\}$ has in turn a further subsequence $\{y_{n(k_l)}\}_{l=1}^{\infty}$ which converges, say to y_0. It follows from Theorem 2.2 that the subsequence $\{z_{n(k_l)}\}_{l=1}^{\infty}$ converges to $z_0 = x_0 + iy_0$. By Exercise 2.4.1, we saw that z_0 is a limit point of S. The proof is complete.

Corollary 2.2 Every bounded sequence of complex numbers has a convergent subsequence.

Let S be a subset of **C**. A class $\{G_\lambda : \lambda \in \Lambda\}$ of open subsets of **C** is called an *open covering* of S if

$$S \subset \bigcup_{\lambda \in \Lambda} G_\lambda$$

A subclass of an open covering of S which is itself an open covering of S is called a *subcovering* of S.

The following theorem of Heine and Borel (Eduard Heine, 1821–1881, German and Émile Borel, 1871–1956, French) is an extension of a corresponding result in **R**.

Theorem 2.8. *Heine-Borel Covering Theorem* Let S be a compact subset of **C**. (That is, S is closed and bounded.) Then any open covering $\{G_\lambda : \lambda \in \Lambda\}$ of S has a finite subcovering $\{G_{\lambda_j} : j = 1, 2, \ldots, n\}$.

FÉLIX ÉDOUARD ÉMILE BOREL

Émile Borel was born in Sain-Affrique, France, on January 7, 1871. His father was a Protestant village pastor and his mother came from a family of merchants. In 1889 Borel entered first in the École Normale. After graduation he taught at the University of Lille, the École Normale, and the Sorbonne. Borel contributed significantly to analysis and probability theory. He is also considered to be the inventor of game theory. He wrote more than 300 scientific papers, the most important being his thesis; one of the famous results in his thesis is the so-called Heine-Borel theorem. Borel was the immediate predecessor of Lebesgue in the development of measure theory.

Borel married Marguerite Appel in 1901, but they had no children. From 1924 to 1940 Borel was heavily involved in politics, and in 1940 he was imprisoned for a short while by the Germans.

In 1955, while Borel was returning from a scientific meeting in Brazil, he was injured in a fall on the ship. He died in Paris the following year, on February 3, 1956, at the age of 85.

Proof. Suppose there exists an open covering $\{G_\lambda: \lambda \in \Lambda\}$ with no finite subcovering. Since S is bounded, it is contained in some closed square K whose sides have length l. Subdivide K into four closed squares with sides of length $l/2$. Then for at least one of the four squares, say K_1, the set $S \cap K_1$ is not covered by a finite number of sets from $\{G_\lambda: \lambda \in \Lambda\}$. Subdivide K_1 into four squares with sides of length $l/2^2$. Then for at least one of these four squares, say K_2, the set $S \cap K_2$ is not covered by a finite number of sets from $\{G_\lambda: \lambda \in \Lambda\}$. Continuing this process, we obtain a sequence

$$S \cap K_1 \supset S \cap K_2 \supset \cdots \supset S \cap K_n \supset \cdots$$

of nonempty closed subsets of $\mathbf{C}$ with

$$\lim_{n \to \infty} d(S \cap K_n) \leq \lim_{n \to \infty} \frac{l}{2^n} = 0$$

EDUARD HEINE

Eduard Heine was born in Berlin, Germany, on March 16, 1821, the eighth of nine children. His father was a banker. Eduard studied at Berlin and Göttingen and from 1848 was a professor at the University of Halle. He was still teaching there when Georg Cantor joined the faculty in 1874.

Influenced by Weierstrass's lectures at Göttingen, Heine introduced the ε-δ definition of limits. He published about 50 papers, most of them dealing with special functions, but his name is best known for its association with the so-called Heine-Borel theorem, which was actually stated and proved by Borel.

Heine married Sophie Wolff in 1850; four daughters and one son were born of this marriage. He died in Halle on October 21, 1881, at the age of 60.

HENRI LÉON LEBESGUE

Henri Lebesgue was born in Beauvais, France, on June 28, 1875. He received his Ph.D. from the University of Paris in 1897. He taught at various universities and in 1921 became a professor at the Collège de France. Although Lebesgue was a brilliant and creative mathematician, honored both at home and abroad, he chose to specialize in just one topic. Today he is regarded as the founder of the modern theory of integration. He died in Paris on July 26, 1941.

Also if $n \in \mathbf{N}$, no finite subclass of $\{G_\lambda : \lambda \in \Lambda\}$ covers $S \cap K_n$. By Theorem 2.6, there exists a point $z_0 \in \bigcap_{n=1}^{\infty} (S \cap K_n)$. In particular, $z_0 \in S$ and so there exists $\lambda_0 \in \Lambda$ such that $z_0 \in G_{\lambda_0}$. As G_{λ_0} is open, there exists an ε neighborhood $N_\varepsilon(z_0) \subset G_{\lambda_0}$. Since $\lim_{n \to \infty} d(K_n) = 0$ and $z_0 \in K_n$ for all $n \in \mathbf{N}$, it follows that there exists an $n_0 \in \mathbf{N}$ such that $K_{n_0} \subset N_\varepsilon(z_0) \subset G_{\lambda_0}$. Hence G_{λ_0}, alone, covers $S \cap K_{n_0}$ and this contradiction proves the theorem.

We finally prove the following result of Henri Lebesgue (1875–1941, French).

Theorem 2.9. Lebesgue's Covering Lemma $\big|$ Let $\{G_\lambda : \lambda \in \Lambda\}$ be an open covering of a compact set S in $\mathbf{C}$. Then there exists $\varepsilon > 0$ such that for every $z_0 \in S$, the ε neighborhood $N_\varepsilon(z_0)$ is contained in at least one of the sets G_λ. (The number ε is called a *Lebesgue number* for the open covering $\{G_\lambda : \lambda \in \Lambda\}$.)

Proof. Since $\{G_\lambda : \lambda \in \Lambda\}$ is an open covering of S, we see that for each $z \in S$, there exists $\lambda(z) \in \Lambda$ and $\varepsilon(z) > 0$ such that $N_{2\varepsilon(z)}(z) \subset G_{\lambda(z)}$. So because $\{N_{\varepsilon(z)}(z) : z \in S\}$ is an open covering of S and S is compact, there exist points $z_1, z_2, \ldots, z_n \in S$ such that $S \subset \bigcup_{k=1}^{n} N_{\varepsilon(z_k)}(z_k)$. Set

$$\varepsilon = \min \{\varepsilon_{z(1)}, \varepsilon_{z(2)}, \ldots, \varepsilon_{z(n)}\}$$

and let $z_0 \in S$. It suffices to show that there exists $\lambda \in \Lambda$ such that $N_\varepsilon(z_0) \subset G_\lambda$. Clearly there exists $k \in \mathbf{N}$, $1 \leq k \leq n$, such that $z_0 \in N_{\varepsilon(z_k)}(z_k)$. Now for any $z \in N_\varepsilon(z_0)$, $d(z_k, z) \leq d(z_k, z_0) + d(z_0, z) < \varepsilon_{z(k)} + \varepsilon \leq 2\varepsilon_{z(k)}$. Hence $N_\varepsilon(z_0) \subset N_{2\varepsilon(z_k)}(z_k) \subset G_{\lambda(z_k)}$ and the proof is complete.

As a corollary, we have the following.

Corollary 2.3 Let $\{G_\lambda : \lambda \in \Lambda\}$ be an open covering of the closed interval $[a,b]$. Then there exist real numbers $a = t_0 < t_1 < \cdots < t_n = b$ such that for each k, $1 \leq k \leq n$, there exists $\lambda \in \Lambda$ such that $[t_{k-1}, t_k] \subset G_\lambda$.

EXERCISES

2.5.1 Show that a subset S of $\mathbf{C}$ is compact if and only if every sequence in S has a convergent subsequence whose limit is in S.

2.5.2 Let A and B be nonempty closed subsets of $\mathbf{C}$ and let A be bounded. Then there exist points $a \in A$ and $b \in B$ such that $d(A,B) = |a - b|$.

2.5.3 Let $S \subset \mathbf{C}$ be compact, and $f: S \to \mathbf{C}$ be continuous. Show that f is uniformly continuous. That is, a continuous function on a compact set is uniformly continuous.

2.5.4 Let $S \subset \mathbf{C}$ be compact. Let $f: S \to \mathbf{C}$ be continuous. Show that $f[S]$ is compact. That is, the continuous image of a compact set is compact.

ARCWISE CONNECTED SETS AND DOMAINS

Let $a \leq b$, and let $\gamma \colon [a,b] \to \mathbf{C}$ be a continuous function. Then we call γ a *curve*. If $A \subset \mathbf{C}$ and im $\gamma \subset A$, then we say γ is a *curve in* A. A set S is called *arcwise connected* if every pair of points in S can be joined by a curve that lies in S. An open and arcwise connected set S is called a *domain*. The sets S and D in Fig. 2.3 are examples of domains.

Let z_1 and z_2 be complex numbers. The *line segment* connecting z_1 to z_2 is the curve denoted by $\overline{z_1 z_2}$ and defined by

$$\overline{z_1 z_2}(t) = z_1 + (z_2 - z_1)t \qquad \text{for } 0 \leq t \leq 1$$

Let $z_0, z_1, \ldots, z_n$ be $n + 1$ points in $\mathbf{C}$ ($n \geq 1$). The n line segments $\overline{z_1 z_2}, \overline{z_2 z_3}, \ldots, \overline{z_{n-1} z_n}$ taken in this order form a curve which is called a *polygonal curve* joining z_0 to z_n with vertices $z_0, \ldots, z_n$; it is denoted by $\overline{z_0 z_1 \cdots z_n}$. To be more precise, $\overline{z_0 z_1 \cdots z_n} \colon [0,n] \to \mathbf{C}$ such that for each $j = 0, 1, \ldots, n - 1$,

$$\overline{z_0 z_1 \cdots z_n}(t) = z_j + (z_{j+1} - z_j)(t - j) \qquad j \leq t \leq j + 1 \ .$$

The points z_0 and z_n are called the *initial* and *terminal* points of the polygonal curve, respectively.

A domain D is called a *star domain* if there is a point $z_0 \in D$ such that $z \in D$ implies that the line segment $\overline{z_0 z}$ lies in D. When the role of z_0 must be emphasized, it is said that D is a star domain with respect to z_0. Clearly the domain S in Fig. 2.3 is not a star domain. The interior of a circle or triangle is a star domain. On the other hand, by deleting three radial segments from an open disk but retaining the center, we obtain a domain D that is a star domain with respect to the center. See Fig. 2.3.

Fig. 2.3

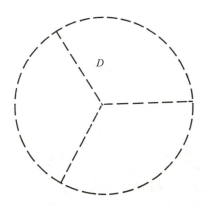
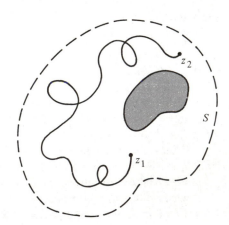

EXERCISES

2.6.1 Let $f: S \to \mathbf{C}$ be continuous and let S be arcwise connected. Show that $f[S]$ is arcwise connected. That is, the continuous image of an arcwise connected set is arcwise connected.

2.6.2 Let D_1 and D_2 be star domains with respect to the same point z_0. Show that $D_1 \cup D_2$ and $D_1 \cap D_2$ are also star domains with respect to z_0.

2.6.3 Classify the following sets according to the terms open, closed, bounded arcwise connected, domain, and star domain. Find also their boundaries, limit points, isolated points, closures, and interiors.

a. $|z - 3| < 1$
b. $|\operatorname{Re} z| \le 1$
c. $|\operatorname{Im} z| < 2$
d. $0 < |z - i| \le 1$
e. $|z + i| + |z - i| = 1$
f. $|z - 2| = |z|$
g. $0 < |1 - z| < 1 + |z|$
h. $1 < \operatorname{Im}(z - 1) < 2$
i. $0 < |z| < 1$
j. $\operatorname{Re} z^2 \ge 0$
k. $0 < \arg z < \pi/6$
l. $\operatorname{Re}(z - iz) \ge 2$
m. $\{z \in \mathbf{C}: |z| < 4\} - (\{z \in \mathbf{C}: |z - i| < 1\} \cup \{z \in \mathbf{C}: |z + i| < 1\})$
n. $\{z \in \mathbf{C}: z = i^n[(ni)/(n + 1)] \text{ for some } n \in \mathbf{N}\}$

2.7

THE EXTENDED COMPLEX PLANE

It is sometimes convenient to extend the system $\mathbf{C}$ of complex numbers by joining to it an "ideal" point denoted by ∞ and called the *point at infinity*. The new set $\mathbf{C} \cup \{\infty\}$, denoted by $\tilde{\mathbf{C}}$ is called the *extended* complex plane.

In order to define the concept of convergence in $\tilde{\mathbf{C}}$, we need the definition of an ε neighborhood of ∞. If $\varepsilon > 0$, an ε *neighborhood of* ∞ is the set $N_\varepsilon(\infty) = \{z: |z| > 1/\varepsilon\} \cup \{\infty\}$. We now extend the concept of convergence as follows: If $z_1, z_2, \ldots$ is a sequence in $\tilde{\mathbf{C}}$ and $z_0 \in \tilde{\mathbf{C}}$, then $\lim_{n \to \infty} z_n = z_0$ if $\varepsilon > 0$ implies that there exists an integer $n_0 > 0$ such that if $n \ge n_0$, then $z_n \in N_\varepsilon(z_0)$. Thus if $\lim_{n \to \infty} z_n = z_0 \in \mathbf{C}$, then there exists a positive integer n_0 such that if $n \ge n_0$, then $z_n \in \mathbf{C}$. On the other hand, if $z_1, z_2, \ldots$ is a sequence in $\mathbf{C}$, then $\lim_{n \to \infty} z_n = \infty \in \tilde{\mathbf{C}}$ is equivalent to the real limit $\lim_{n \to \infty} |z_n| = +\infty$.

Hence if $z_1, z_2, \ldots$ and $\zeta_1, \zeta_2, \ldots$ are sequences in $\mathbf{C}$ such that $\lim_{n \to \infty} z_n = \infty$ and $\lim_{n \to \infty} \zeta_n = a \in \mathbf{C}$, it follows that

$$\lim_{n \to \infty} (\zeta_n + z_n) = \lim_{n \to \infty} (z_n + \zeta_n) = \infty$$

This property of sequences justifies the algebraic operation

$$a + \infty = \infty + a = \infty \qquad a \in \mathbf{C}$$

In a similar way the following operations with ∞ are justified:

$$\frac{a}{\infty} = 0 \qquad\qquad a \in \mathbf{C}$$

$$a\infty = \infty a = \infty \qquad a \in \mathbf{C} - \{0\}$$

$$\frac{a}{0} = \infty \qquad\qquad a \in \mathbf{C} - \{0\}$$

Since $\lim_{n \to \infty} [n + (-n)] = 0$ and $\lim_{n \to \infty} (n + n) = \infty$, we see that no definition of $\infty + \infty$ can be made. Similarly $-\infty + \infty$, $-\infty - \infty$, $0 \cdot \infty$, $\infty \cdot 0$, ∞/∞, and $0/0$ are all undefined.

The other definitions found in the preceding sections of this chapter are similarly extended. Thus if $A \subset \tilde{\mathbf{C}}$ and $z_0 \in \tilde{\mathbf{C}}$, then z_0 is a limit point of A if $\varepsilon > 0$ implies $[A \cap N_\varepsilon(z_0)] - \{z_0\} \neq \varnothing$. So we see that a set $A \subset \mathbf{C}$ is unbounded if and only if ∞ is a limit point of A in $\tilde{\mathbf{C}}$. It follows that every infinite set S in $\tilde{\mathbf{C}}$ has a limit point in $\tilde{\mathbf{C}}$. In fact, if S contains an infinite bounded subset S^*, the existence of a limit point follows from Theorem 2.7 (the Bolzano-Weierstrass theorem) applied to S^*, whereas if S does not contain any infinite bounded subsets, then every ε neighborhood of ∞ contains points of S different from ∞, and so ∞ itself is a limit point of S. Thus we see that the Bolzano-Weierstrass theorem for $\tilde{\mathbf{C}}$ does not require the "boundedness" hypothesis of S which is so essential for the theorem for $\mathbf{C}$. For topological spaces more general than $\mathbf{C}$, a set is said to be *compact* if every open cover has a finite subcover. Using the Bolzano-Weierstrass property for $\tilde{\mathbf{C}}$ and our new definition for compactness, one can easily show that $\tilde{\mathbf{C}}$ is compact.

We shall next give a geometric representation of $\tilde{\mathbf{C}}$ which will at the same time give us a new geometric representation of $\mathbf{C}$. Consider the unit sphere

$$\Sigma = \{(\xi,\eta,\zeta) \in \mathbf{R}^3 : \xi^2 + \eta^2 + \zeta^2 = 1\}$$

in $\mathbf{R}^3$. The equator plane

$$\Pi = \{(\xi,\eta,0) : \xi, \eta \in \mathbf{R}\}$$

can be taken to be $\mathbf{C}$, and in fact if $(\xi,\eta,0) \in \Pi$, we shall write (ξ,η) instead of $(\xi,\eta,0)$. See Fig. 2.4.

The sphere Σ is called the *Riemann sphere*, and the point $N = (0,0,1)$ is called the *north pole* of Σ. The line joining a point z of the equator plane to the north pole N intersects Σ in a unique point $P_z \neq N$. Conversely, the line joining N with a point P_z of Σ different from N intersects the plane Π in a unique point z. It is clear that this construction establishes a one-to-one correspondence between the points of the equator plane (the complex plane) and $\Sigma^* = \Sigma - \{N\}$. This mapping T of Σ^* onto Π is called *stereographic projection*. We now formulate the analytic description of this mapping. Let $P_z = (\xi,\eta,\zeta) \in \Sigma^*$. The equation of the line joining N to P_z is

Fig. 2.4

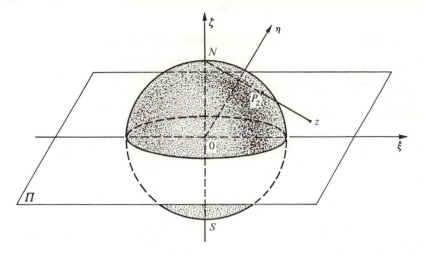

$$\frac{X - \xi}{\xi} = \frac{Y - \eta}{\eta} = \frac{Z - \zeta}{\zeta}$$

When $Z = 0$, this line intersects Π at the point $z = (x, y)$ where

$$x = \frac{\xi}{1 - \zeta} \quad \text{and} \quad y = \frac{\eta}{1 - \zeta}$$

Hence we see that the mapping $T: \Sigma^* \to \Pi$ is given by

$$T(\xi, \eta, \zeta) = \left(\frac{\xi}{1 - \zeta}, \frac{\eta}{1 - \zeta} \right)$$

It is not difficult to show that T is one-to-one and onto, and in fact, one can show that

$$T^{-1}: \Pi \to \Sigma^*$$

is given by

$$T^{-1}(x, y) = \left(\frac{2x}{x^2 + y^2 + 1}, \frac{2y}{x^2 + y^2 + 1}, \frac{x^2 + y^2 - 1}{x^2 + y^2 + 1} \right)$$

In complex notation,

$$T: \Sigma^* \to \mathbf{C}$$

is given by

$$T(\xi, \eta, \zeta) = \frac{\xi + i\eta}{1 - \zeta}$$

and

$$T^{-1}: \mathbf{C} \to \Sigma^*$$

is given by

$$T^{-1}(z) = \left(\frac{z + \bar{z}}{|z|^2 + 1}, \frac{z - \bar{z}}{i(|z|^2 + 1)}, \frac{|z|^2 - 1}{|z|^2 + 1} \right)$$

We leave to the reader the task of showing that T and T^{-1} are continuous, and hence that T is a *homeomorphism* of Σ^* onto C. Consequently, we may consider the set of all complex numbers as points on the punctured sphere Σ^*, rather than as points on a plane. In this way we have another geometric interpretation of the complex numbers. If in addition we extend the above mapping T to $\tilde{T}: \Sigma \to \tilde{C}$ by setting $\tilde{T}(N) = \infty$, then $\tilde{T}$ is a homeomorphism of Σ onto $\tilde{C}$. (In other words, $\tilde{T}: \Sigma \to \tilde{C}$ is one-to-one and onto; $\mathcal{O}$ is open in $\tilde{C}$ implies $\tilde{T}^{-1}[\mathcal{O}]$ is open in Σ, and V is open in Σ implies $\tilde{T}[V]$ is open in $\tilde{C}$.) Thus we can regard the Riemann sphere as a representation of $\tilde{C}$. Notice that the hemisphere $\zeta < 0$ in Σ corresponds to the disk $|z| < 1$, while the hemisphere $\zeta > 0$ corresponds to the set $\{z: |z| > 1\} \cup \{\infty\} = N_1(\infty)$.

In the spherical representation of $\tilde{C}$, there is no simple interpretation of addition and multiplication; its advantage lies in the fact that the point at infinity is no longer an "ideal" point.

A characteristic property of stereographic projection is the following:

Every circle in the complex plane corresponds to a circle on Σ missing the north pole N, and every straight line in the complex plane corresponds to a circle on Σ passing through N. Conversely, every circle on Σ is projected onto a circle or a straight line in the complex plane, according to whether it passes through the north pole or not.

If a straight line is regarded as a special kind of circle, namely, a "circle through the point at infinity," we then have the following theorem.

Theorem 2.10 Stereographic projection preserves circles.

Proof. Let

$$a(x^2 + y^2) + bx + cy + d = 0 \qquad a^2 + b^2 + c^2 \neq 0 \qquad (2.15)$$

be the equation of a circle (a straight line if $a = 0$) in the complex plane. Under stereographic projection, we have

$$x = \frac{\xi}{1 - \zeta} \qquad \text{and} \qquad y = \frac{\eta}{1 - \zeta} \qquad \zeta \neq 1$$

Hence Eq. (2.15) becomes

$$a \frac{\xi^2 + \eta^2}{(1 - \zeta)^2} + b \frac{\xi}{1 - \zeta} + c \frac{\eta}{1 - \zeta} + d = 0$$

Since $\xi^2 + \eta^2 + \zeta^2 = 1$, this equation reduces to

$$a(1 + \zeta) + b\xi + c\eta + d(1 - \zeta) = 0 \qquad (2.16)$$

which is the equation of a plane. This plane cuts the sphere Σ in a circle.

If $a = 0$, the plane (2.16) becomes

$$b\xi + c\eta + d(1 - \zeta) = 0$$

which clearly passes through the north pole $N = (0,0,1)$; but if $a \neq 0$, then clearly the plane does not pass through N. Conversely, let

$$A\xi + B\eta + C\zeta + D = 0 \qquad \xi^2 + \eta^2 + \zeta^2 = 1 \qquad (2.17)$$

with $A^2 + B^2 + (C + D)^2 \neq 0$ be the equation of a circle on Σ. [If $C + D = 0$, the circle (2.17) passes through N, but if $C + D \neq 0$, the circle does not pass through N.] Then under stereographic projection [see the formula giving $T^{-1}(x,y)$], we have

$$\xi = \frac{2x}{x^2 + y^2 + 1} \qquad \eta = \frac{2y}{x^2 + y^2 + 1} \qquad \zeta = \frac{x^2 + y^2 - 1}{x^2 + y^2 + 1}$$

Hence after some manipulations, Eq. (2.17) reduces to

$$(C + D)(x^2 + y^2) + 2Ax + 2By + (D - C) = 0 \qquad (2.18)$$

Clearly, if $C + D \neq 0$, then Eq. (2.18) represents a circle in the complex plane, and if $C + D = 0$, Eq. (2.18) reduces to a line. The proof is complete.

Another interesting property of stereographic projection is the fact that "the angle between the tangents of two intersecting differentiable curves on the Riemann sphere Σ is equal to the angle between their stereographic images in the extended complex plane." Cartographers find this property of stereographic projection very useful in preparing navigational maps.

EXERCISES

2.7.1 If a_n and b_m are nonzero, show that

$$\lim_{z \to \infty} \frac{a_n z^n + \cdots + a_1 z + a_0}{b_m z^m + \cdots + b_1 z + b_0} = \begin{cases} \infty & n > m \\ a_n/b_n & n = m \\ 0 & n < m \end{cases}$$

2.7.2 Show that the set $S = \{1, 2, \ldots\}$ is closed in $\mathbf{C}$ but not in $\tilde{\mathbf{C}}$.

2.7.3 Let $\{P_{z_n}\}$ be a sequence on Σ and $\{z_n\}$ the corresponding sequence on the extended complex plane obtained by stereographic projection. (Recall that the stereographic projection of N is ∞.) Show that

$$\lim_{n \to \infty} P_{z_n} = P_z \quad \text{if and only if} \quad \lim_{n \to \infty} z_n = z$$

Chapter 3 Analytic Functions

3.1

INTRODUCTION

Having studied the algebra and topology of complex numbers, we are now ready to consider the differential calculus of functions of a complex variable. In Chapter 2 we stressed the similarities of the topology of complex numbers and complex functions with that of real numbers and real functions. For example, a complex function is continuous if and only if both the real and imaginary parts of the function are continuous. A similar statement concerning the differentiability of complex functions would not be true.

We start this chapter by defining the concept of analytic functions. In order that we may have available a vast class of examples of analytic functions, we begin by studying complex power series. Every convergent power series is an analytic function (and, as we shall see in Chapter 4, every analytic function is locally a power series). With the theory of power series, we are able to extend exponential, trigonometric, and hyperbolic functions in such a way that all algebraic identities concerning these functions in the real case are also true when the functions are considered as functions of a complex variable. The logarithm is also treated here, but unlike the "real case," the logarithm of a complex number is "multiple-valued."

Finally, we investigate the conditions that the differentiability of $f(z) = u(x,y) + iv(x,y)$ imposes on the real-valued functions u and v. This leads us to the *Cauchy-Riemann* equations, which are a pair of first-order partial differential equations involving u and v. In addition, u and v must each satisfy a second-order partial differential equation called *Laplace's partial differential equation* (Pierre Simon, Marquis de Laplace, 1749–1827, French). This fact is the basis of many applications to physical problems.

3.2
THE COMPLEX DERIVATIVE

In the sequel, unless otherwise specified, D will stand for a domain (an open and arcwise connected set) in $\mathbf{C}$.

Let $f: D \to \mathbf{C}$ be a function and let $z_0 \in D$. We say that f is *differentiable* at z_0 if

$$\lim_{z \to z_0} \frac{f(z) - f(z_0)}{z - z_0}$$

exists. This limit, called the *derivative* of f at z_0, is denoted by $f'(z_0)$.

Writing $z - z_0 = h$ in the above definition, we get the equivalent form

$$\lim_{h \to 0} \frac{f(z_0 + h) - f(z_0)}{h}$$

From the identity

$$f(z) = f(z_0) + \frac{f(z) - f(z_0)}{z - z_0}(z - z_0) \qquad z \neq z_0$$

it follows that if f is differentiable at z_0, then

$$\lim_{z \to z_0} f(z) = f(z_0)$$

that is, f is also continuous at z_0. Clearly the converse is not true, as the function

$$f(z) = \bar{z} \qquad z \in \mathbf{C}$$

is continuous but nowhere differentiable (see Example 3.2 below).

In the theory of complex functions, we are especially interested in functions $f: D \to \mathbf{C}$ which possess a derivative not only at a single point $z_0 \in D$, but at every point in an ε neighborhood of z_0.

We say that a function $f: D \to \mathbf{C}$ is *analytic* at a point $z_0 \in D$ if f is differentiable at every point in some ε neighborhood of z_0. If f is analytic at every point of D, then f is called *analytic in D*. An analytic function is also called *holomorphic, monogenic,* or *regular*. The set of all points where a function is analytic is called the *domain of analyticity* of the function. A function whose domain of analyticity is the complex plane $\mathbf{C}$ is called *entire*.

If f is analytic in D, then for every $z \in D$ the derivative $f'(z)$ of f at z exists. This process defines a new function $f': D \to \mathbf{C}$, which is called the *first derivative* of f in D. Higher derivatives of f are defined in an obvious way. Exactly as in the calculus, one can prove the following formulas:

$$(f + g)' = f' + g' \qquad (fg)' = f'g + fg' \qquad \left(\frac{f}{g}\right)' = \frac{f'g - fg'}{g^2}$$

Finite sums and products of analytic functions are also analytic, and the quotient of two analytic functions is analytic wherever the denominator is different from zero. Finally, the composition of two analytic functions is analytic; more precisely, if f is differentiable at z_0 and g is differentiable at $f(z_0)$, then $h = g \circ f$ is differentiable at z_0 and the *chain rule* holds:

$$h'(z_0) = g'(f(z_0))f'(z_0)$$

Clearly any constant function is an entire function with derivative equal to zero. The function $f(z) = z$ is also entire and has derivative $f'(z) = 1$.

Example 3.1 Show that the function

$$f(z) = az^n$$

is an entire function, and that

$$f'(z_0) = anz_0^{n-1} \qquad z_0 \in \mathbf{C}$$

Proof. We have

$$\frac{f(z) - f(z_0)}{z - z_0} = \frac{az^n - az_0^n}{z - z_0} = a(z^{n-1} + z^{n-2}z_0 + z^{n-3}z_0^2 + \cdots + z_0^{n-1})$$

Hence

$$\lim_{z \to z_0} \frac{f(z) - f(z_0)}{z - z_0} = a(\underbrace{z_0^{n-1} + z_0^{n-1} + \cdots + z_0^{n-1}}_{n \text{ terms}}) = anz_0^{n-1}$$

It follows from Example 3.1 and the previous observations that a polynomial is an entire function. Also, the quotient of two polynomials is analytic everywhere except where the denominator is zero.

Example 3.2 Show that the function $f(z) = \bar{z}$ is nowhere differentiable in $\mathbf{C}$.

Proof. Observe that

$$\frac{f(z + h) - f(z)}{h} = \frac{\overline{z + h} - \bar{z}}{h} = \frac{\bar{h}}{h}$$

which approaches 1 as $h \to 0$ through real values and approaches -1 as $h \to 0$ through pure-imaginary values. Thus

$$\lim_{h \to 0} \frac{f(z + h) - f(z)}{h}$$

does not exist, and the proof is complete.

Example 3.3 Show that the function

$$f(z) = |z|^2$$

is differentiable only at zero, and is therefore nowhere analytic.

Proof. If $h \neq 0$, then

$$\frac{f(z + h) - f(z)}{h} = \frac{|z + h|^2 - |z|^2}{h} = z\frac{\bar{h}}{h} + \bar{z} + h \qquad (3.1)$$

If $z = 0$, the limit as $h \to 0$ of the difference quotient in Eq. (3.1) exists and is equal to zero; if $z \neq 0$, then, as in Example 3.2, the limit does not exist. So the function $f(z) = |z|^2$ is differentiable only at $z = 0$. Since f is not differentiable in any neighborhood of 0, f is not analytic at 0 (and clearly not analytic anywhere else).

Example 3.4 Let $f: D \to \mathbf{R}$ be a real-valued function of a complex argument, and let $z_0 \in D$. Then either $f'(z_0) = 0$ or else f is not differentiable at z_0.

Proof. Assume that the derivative of f exists at $z_0 \in D$ and is different from zero. Then as $h \to 0$ through real values,

$$\lim_{h \to 0} \frac{f(z_0 + h) - f(z_0)}{h}$$

is real; and as $h \to 0$ through pure-imaginary values,

$$\lim_{h \to 0} \frac{f(z_0 + h) - f(z_0)}{h}$$

is pure-imaginary. Hence $f'(z_0)$ does not exist. This contradiction proves the result.

EXERCISES

3.2.1 Find the domain of analyticity and compute the derivative of each of the following functions where it exists.

a. $\left(z - \dfrac{1}{z}\right)^3$

b. $\dfrac{z^2 - 4}{z^3 - 3z - 2}$

c. $\dfrac{z + i}{z - 2}$

d. $\dfrac{i}{z^3}$

3.2.2 Show that the functions Re z, Im z, Arg z, and $|z|$ are nowhere analytic in $\mathbf{C}$.

3.2.3 Let $f\colon D \to \mathbf{C}$ take only pure-imaginary values, and let $z_0 \in D$. Show that either $f'(z_0) = 0$ or else f is not differentiable at z_0.

3.2.4 Let $z = x + iy$ and $f\colon \mathbf{C} \to \mathbf{C}$ be defined by

$$
f(z) = \begin{cases} \dfrac{x^2 yz}{x^4 + y^2} & z \neq 0 \\[2mm] 0 & z = 0 \end{cases}
$$

Show that f is not differentiable at $z = 0$.

3.2.5 Show that the functions $z\,\mathrm{Re}\,z$ and $z\,\mathrm{Im}\,z$ are differentiable but not analytic at $z = 0$.

3.2.6 Show that if a polynomial P is of stable type (that is, all its roots have negative real parts), then P' is also of stable type. Show also that if the roots of P have positive real parts, then the same is true for the roots of P'.

3.3
PROPERTIES OF POWER SERIES

Consider the power series which has center z_0 and coefficients $a_0, a_1, \dots$.

$$
\sum_{n=0}^{\infty} a_n(z - z_0)^n = a_0 + a_1(z - z_0) + \cdots + a_n(z - z_0)^n + \cdots \qquad (3.2)
$$

In this section, we develop various properties of the above power series.

Lemma 3.1 If the power series (3.2) converges at some point z_1 and diverges at some point z_2, then it converges absolutely for all z such that $|z - z_0| < |z_1 - z_0|$, and it diverges for all z such that $|z - z_0| > |z_2 - z_0|$.

Proof. Since

$$
a_0 + a_1(z_1 - z_0) + \cdots + a_n(z_1 - z_0)^n + \cdots
$$

converges, its nth term approaches zero as $n \to \infty$, and therefore there exists a positive number M such that

$$
|a_n(z_1 - z_0)^n| \leq M \qquad n = 0, 1, 2, \dots
$$

Suppose that $|z - z_0| < |z_1 - z_0|$. Then

$$
\sum_{n=0}^{\infty} |a_n(z - z_0)^n| = \sum_{n=0}^{\infty} |a_n(z_1 - z_0)^n| \left| \frac{z - z_0}{z_1 - z_0} \right|^n \leq M \sum_{n=0}^{\infty} \omega^n
$$

where $\omega = |(z - z_0)/(z_1 - z_0)| < 1$. It follows that the series (3.2) converges absolutely for $|z - z_0| < |z_1 - z_0|$.

The remaining part of this lemma is proved similarly.

Let

$$R = \frac{1}{\displaystyle\lim_{n \to \infty} \sup_{k \geq n} \sqrt[k]{|a_k|}}$$

where $a_0, a_1, a_2, \ldots$ are the coefficients of the power series (3.2). R is called the *radius of convergence* of (3.2). The following result of Cauchy and Hadamard (Jacques Hadamard, 1865–1963, French) is very useful.

Lemma 3.2. The Cauchy-Hadamard Rule If $R = 0$, series (3.2) converges only for $z = z_0$. If $R = \infty$, series (3.2) converges absolutely for all $z \in \mathbf{C}$. If $0 < R < \infty$, series (3.2) converges absolutely if $|z - z_0| < R$, and it diverges if $|z - z_0| > R$.

Proof. Let us set

$$\alpha_n = a_n(z - z_0)^n$$

Then

$$\lim_{n \to \infty} \sup_{k \geq n} \sqrt[k]{|\alpha_k|} = |z - z_0| \lim_{n \to \infty} \sup_{k \geq n} \sqrt[k]{|a_k|} = \frac{|z - z_0|}{R}$$

By Theorem 2.5, the power series converges absolutely if $|z - z_0|/R < 1$, and it diverges if $|z - z_0|/R > 1$. The cases $R = 0$ and $R = \infty$ are similar and are left to the reader.

When $0 < R < \infty$, the circle $|z - z_0| = R$ is called the *circle of convergence* of the power series (3.2).

JACQUES SALAMON HADAMARD

Jacques Hadamard was born at Versailles, France, on December 8, 1865. His father was a Latin teacher and his mother was a piano teacher. He entered the École Normale in 1884 and received his D.Sc. in 1892. Hadamard taught as professor at the Collège de France from 1909 to 1937 and at the École Polytechnique from 1912 to 1937. He made significant contributions to complex variables, number theory, and differential equations, and his work greatly influenced many branches of applied mathematics.

Hadamard died in Paris on October 17, 1963.

Lemma 3.2 states that a power series converges absolutely at every point in the interior of its circle of convergence, and it diverges at every point exterior to its circle of convergence. In general, nothing can be said regarding convergence on the circle of convergence. It may or may not converge there. Also, it may converge at some points and diverge at others. We illustrate this by the following example.

Example 3.5 Show that (a) $\sum_{n=0}^{\infty} n z^n$ diverges at all points of its circle of convergence; (b) $\sum_{n=1}^{\infty} z^n/n^3$ converges at all points of its circle of convergence; (c) $\sum_{n=1}^{\infty} z^n/\sqrt{n}$ converges at $z = -1$ and diverges at $z = 1$.

Proof

a. Recall that $\lim_{n \to \infty} \sqrt[n]{n} = 1$. So clearly, $R = 1$ and the circle of convergence is the set $|z| = 1$. The series in (a) clearly diverges for all z with $|z| = 1$, since the nth term does not even approach zero as $n \to \infty$.
b. The circle of convergence is the set $|z| = 1$. If $|z| = 1$, $|z^n/n^3| = 1/n^3$. Since $\sum_{n=1}^{\infty} 1/n^3$ converges, the result is clear.
c. The circle of convergence is again the set $|z| = 1$. However, $\sum_{n=1}^{\infty} (-1)^n/\sqrt{n}$ converges, whereas $\sum_{n=1}^{\infty} 1/\sqrt{n}$ diverges.

Let us break up the power series (3.2) as follows:

$$\sum_{n=0}^{\infty} a_n(z - z_0)^n = \sum_{n=0}^{N-1} a_n(z - z_0)^n + R_N(z)$$

where $R_N(z) = \sum_{n=N}^{\infty} a_n(z - z_0)^n$ is called the *remainder* of the series after N terms. We also define

$$S_{N-1}(z) = \sum_{n=0}^{N-1} a_n(z - z_0)^n$$

We say that the series (3.2) *converges uniformly* in a set S if for each $\varepsilon > 0$, there exists a positive integer $N(\varepsilon)$ such that for all $z \in S$ and $N \geq N(\varepsilon)$,

$$|R_N(z)| < \varepsilon$$

Lemma 3.3 Let $R > 0$ be the radius of convergence of the power series (3.2). Let $0 < r < R$. Then (3.2) converges uniformly in the set $|z - z_0| \leq r$.

Proof. Let $\varepsilon > 0$. Let $r < R_1 < R$. Then there exists ω with $0 < \omega < 1$ such that $r = \omega R_1$. If $|z - z_0| = R_1$, then $\sum_{n=0}^{\infty} a_n(z - z_0)^n$ converges absolutely, and so there exists an integer N_1 such that $n \geq N_1$ implies $|a_n| R_1^n < 1$. Also since $0 < \omega < 1$, there exists $N_2 \in \mathbf{N}$ such that

$$\omega^n < \varepsilon(1 - \omega) \qquad n \geq N_2$$

Let $N(\varepsilon) = \max (N_1, N_2)$. Then if $N \geq N(\varepsilon)$ and $|z - z_0| \leq r$,

$$|R_N(z)| \leq \sum_{n=N}^{\infty} |a_n(z - z_0)^n| \quad \text{(by Theorem 2.4)}$$

$$\leq \sum_{n=N}^{\infty} |a_n|\omega^n R_1^n < \sum_{n=N}^{\infty} \omega^n = \frac{\omega^N}{1 - \omega} < \varepsilon$$

and the proof is complete.

When the power series (3.2) converges to a complex number $f(z)$ for each point z in a set S, we say that the series *represents* the function f in S (or that the function f is a power series on S), and we write

$$f(z) = \sum_{n=0}^{\infty} a_n(z - z_0)^n \qquad z \in S$$

Theorem 3.1 In the interior of its circle of convergence, the power series

$$f(z) = \sum_{n=0}^{\infty} a_n(z - z_0)^n$$

is an analytic function. Moreover,

$$f'(z) = \sum_{n=0}^{\infty} n a_n(z - z_0)^{n-1}$$

That is, the derivative of f can be found by term-by-term differentiation. Furthermore, the radius of convergence of f' is the radius of convergence of f.

Proof. For simplicity, we assume $z_0 = 0$ and consider the power series

$$g(z) = \sum_{n=0}^{\infty} n a_n z^{n-1}$$

By Lemma 3.2 it follows that g has the same radius of convergence R as f. Let $|z| = r < R$. It suffices to show that $f'(z) = g(z)$. Let $h \neq 0$ be a complex number such that $r + |h| < \rho < R$. Set $|h| = \varepsilon$. Then by the binomial theorem, we have

$$\left| \frac{(z + h)^n - z^n}{h} - nz^{n-1} \right| = \left| \frac{n(n - 1)}{2!} z^{n-2}h + \cdots + h^{n-1} \right|$$

$$\leq \frac{n(n - 1)}{2!} r^{n-2}\varepsilon + \cdots + \varepsilon^{n-1}$$

$$= \frac{(r + \varepsilon)^n - r^n}{\varepsilon} - nr^{n-1}$$

Since the series $\sum_{n=0}^{\infty} |a_n|\rho^n$ converges, there exists a constant K such that $|a_n|\rho^n \leq K$ for $n = 0, 1, 2, \ldots$. Thus

$$\left| \frac{f(z+h) - f(z)}{h} - g(z) \right| = \left| \sum_{n=0}^{\infty} a_n \left[\frac{(z+h)^n - z^n}{h} - nz^{n-1} \right] \right|$$

$$\leq \sum_{n=0}^{\infty} |a_n| \left[\frac{(r+\varepsilon)^n - r^n}{\varepsilon} - nr^{n-1} \right]$$

$$= \sum_{n=0}^{\infty} |a_n|\rho^n \left\{ \frac{[(r+\varepsilon)/\rho]^n - (r/\rho)^n}{\varepsilon} - \frac{n}{r}\left(\frac{r}{\rho}\right)^n \right\}$$

$$\leq K \sum_{n=0}^{\infty} \left\{ \frac{1}{\varepsilon}\left[\left(\frac{r+\varepsilon}{\rho}\right)^n - \left(\frac{r}{\rho}\right)^n \right] - \frac{n}{r}\left(\frac{r}{\rho}\right)^n \right\}$$

Note by Theorem 2.3 that, given $|\omega| < 1$,

$$\sum_{n=0}^{\infty} n\omega^n = \omega + \omega^2 + \omega^3 + \omega^4 + \cdots$$
$$+ \omega^2 + \omega^3 + \omega^4 + \cdots$$
$$+ \omega^3 + \omega^4 + \cdots$$
$$\cdots$$

$$= \frac{\omega}{1-\omega} + \frac{\omega^2}{1-\omega} + \frac{\omega^3}{1-\omega} + \cdots$$

$$= \frac{1}{1-\omega}\frac{\omega}{1-\omega} = \frac{\omega}{(1-\omega)^2}$$

Hence

$$\left| \frac{f(z+h) - f(z)}{h} - g(z) \right|$$

$$\leq K \left\{ \frac{1}{\varepsilon}\left[\frac{1}{1-(r+\varepsilon)/\rho} - \frac{1}{1-(r/\rho)} \right] - \frac{r/\rho}{r[1-(r/\rho)]^2} \right\}$$

$$= K \left[\frac{1}{\varepsilon}\left(\frac{\rho}{\rho-r-\varepsilon} - \frac{\rho}{\rho-r} \right) - \frac{\rho}{(\rho-r)^2} \right]$$

$$= \frac{\rho\varepsilon K}{(\rho-r-\varepsilon)(\rho-r)^2}$$

So as $\varepsilon = |h|$, clearly

$$\lim_{h\to 0} \left| \frac{f(z+h) - f(z)}{h} - g(z) \right| = 0$$

which implies that $f'(z) = g(z)$, and so the proof is complete.

The following two corollaries are elementary consequences of Theorem 3.1.

Corollary 3.1 In the interior of its circle of convergence, a power series is infinitely differentiable. (In particular, it is a continuous function.)

Corollary 3.2 In the interior of its circle of convergence, the power series

$$f(z) = \sum_{n=0}^{\infty} a_n(z - z_0)^n$$

has a "primitive"; in fact, the power series

$$g(z) = \sum_{n=1}^{\infty} \frac{a_{n-1}}{n} (z - z_0)^n$$

has the same radius of convergence as f, and satisfies

$$g'(z) = f(z)$$

The following theorem is of fundamental importance.

Theorem 3.2 Let

$$f(z) = \sum_{n=0}^{\infty} a_n(z - z_0)^n$$

be a power series. Let $\{z_n\}_{n=1}^{\infty}$ be a sequence which converges to z_0 such that $z_n \neq z_0$ for $n = 1, 2, \ldots$. Furthermore assume that $f(z_n) = 0$ for $n = 1, 2, \ldots$. Then $a_n = 0$ for $n = 0, 1, 2, \ldots$.

Proof. As f is continuous, we have

$$a_0 = f(z_0) = f(\lim_{n \to \infty} z_n) = \lim_{n \to \infty} f(z_n) = 0$$

Consider the function

$$g(z) = \sum_{n=1}^{\infty} a_n(z - z_0)^{n-1}$$

Then g is a power series with the same radius of convergence as f, and

$$0 = f(z_n) = g(z_n)(z_n - z_0) \qquad n = 1, 2, \ldots$$

Since $z_n - z_0 \neq 0$, it follows that $g(z_n) = 0$ for $n = 1, 2, \ldots$. Hence $0 = g(z_0) = a_1$. Continuing in this fashion, we prove the theorem.

Corollary 3.3 Let $S \subset \mathcal{C}$ be a set with a limit point z_0 (for example, a real interval), and let

$$\sum_{n=0}^{\infty} a_n(z - z_0)^n = \sum_{n=0}^{\infty} b_n(z - z_0)^n$$

for all $z \in S$. Then $a_n = b_n$ for $n = 0, 2, 1, \ldots$.

EXERCISES

3.3.1 Find the radius of convergence of each of the following power series.

a. $\displaystyle\sum_{n=1}^{\infty} \frac{(-1)^{n-1}}{2n-1}(z-i)^n$

b. $\displaystyle\sum_{n=0}^{\infty} \frac{(z-2+i)^n}{n!}$

c. $\displaystyle\sum_{n=0}^{\infty} \frac{(-1)^n}{(2n+1)!} z^{2n+1}$

d. $\displaystyle\sum_{n=0}^{\infty} \frac{(-1)^n}{(2n)!} z^{2n}$

e. $\displaystyle\sum_{n=1}^{\infty} \frac{(z+i)^n}{(3n)^{1/2}}$

f. $\displaystyle\sum_{n=1}^{\infty} \left(1 + \frac{1}{n}\right)^{n^2} i^n (z-1)^n$

3.3.2 Find the radius of convergence of each of the following power series. Determine for each power series if there are any points on its circle of convergence where the series converges, and anywhere it diverges.

a. $\displaystyle\sum_{n=0}^{\infty} \frac{z^n}{n+1}$

b. $\displaystyle\sum_{n=0}^{\infty} n^{1/2}(z-i)^n$

c. $\displaystyle\sum_{n=2}^{\infty} \frac{(-1)^n}{n(n-1)}(z+2i)^n$

d. $\displaystyle\sum_{n=1}^{\infty} z^{2^n}$

3.3.3 Prove that the series $\sum_{n=0}^{\infty} z^n$ is not uniformly convergent on the set $|z| < 1$.

3.3.4 Prove that the series $\sum_{n=1}^{\infty} z^n/n^4$ is uniformly convergent on the set $|z| \leq 0.9$.

3.3.5 Show that if $|z - 1| < 1$, then

$$\frac{1}{z} = \sum_{n=0}^{\infty} (-1)^n (z-1)^n$$

Using this, determine a power series representing the function $f \colon \{z \in \mathbf{C} \colon |z - 1| < 1\} \to \mathbf{C}$ defined by $f(z) = 1/z^2$.

3.4
EXPONENTIAL AND TRIGONOMETRIC FUNCTIONS

In this section, we extend the definitions of many of the functions that we met in the calculus. The following argument provides the motivation for this section.

Let $a_0, a_1, a_2, \ldots$ be a sequence of real numbers, and let $x_0 \in \mathbf{R}$. Suppose the radius of convergence R of the real power series

$$f(x) = \sum_{n=0}^{\infty} a_n(x - x_0)^n$$

is not zero. We wish to extend the domain of definition of f to make it a complex power series. By Lemma 3.2, the complex power series

$$F(z) = \sum_{n=0}^{\infty} a_n(z - x_0)^n$$

has the same radius of convergence as f and is an obvious candidate for its extension. If

$$G(z) = \sum_{n=0}^{\infty} b_n(z - x_0)^n$$

is any other extension of f, then $F(x) = G(x)$ for all x in some open real interval about x_0, and so by Corollary 3.3, we see that $a_n = b_n$ for $n = 0, 1, 2, \ldots$. Hence

$$F(z) = \sum_{n=0}^{\infty} a_n(z - x_0)^n$$

is the only power series centered at x_0 that extends f. (In Sec. 4.5 we shall prove that an analytic function is locally expressible as a power series, and hence F is actually the only analytic extension of f possible throughout the set $|z - x_0| < R$.) Thus we define the *complex exponential function* by

$$\exp z = \sum_{n=0}^{\infty} \frac{z^n}{n!} \qquad \text{for } z \in \mathbf{C}$$

the *complex sine function* by

$$\sin z = \sum_{n=0}^{\infty} (-1)^n \frac{z^{2n+1}}{(2n+1)!} \qquad \text{for } z \in \mathbf{C}$$

and the *complex cosine function* by

$$\cos z = \sum_{n=0}^{\infty} (-1)^n \frac{z^{2n}}{(2n)!} \qquad \text{for } z \in \mathbf{C}$$

As exponential, sine, and cosine have $R = \infty$ when they are considered as real power series, they also have $R = \infty$ when considered as complex power series. Adding the power series $\cos z$ to the power series $i \sin z$ yields the identity

$$e^{iz} = \cos z + i \sin z \qquad \text{for } z \in \mathbf{C}$$

Similarly, we have *Euler's formulas*:

$$\cos z = \frac{e^{iz} + e^{-iz}}{2} \qquad \text{and} \qquad \sin z = \frac{e^{iz} - e^{-iz}}{2i}$$

We now derive some interesting properties of exp, sin, and cos.

1. $e^0 = 1$.
2. $\exp' z = \exp z$ (where $\exp'$ denotes the derivative of exp).
3. $e^{z+w} = e^z e^w$. In fact,

$$e^z e^w = \sum_{j=0}^{\infty} \frac{z^j}{j!} \sum_{k=0}^{\infty} \frac{w^k}{k!}$$

$$= \sum_{n=0}^{\infty} \sum_{j=0}^{n} \frac{z^j w^{n-j}}{j!\,(n-j)!} \qquad \text{(by Exercise 2.4.8)}$$

$$= \sum_{n=0}^{\infty} \frac{1}{n!} \sum_{j=0}^{n} \binom{n}{j} z^j w^{n-j} = \sum_{n=0}^{\infty} \frac{(z+w)^n}{n!} = e^{z+w}$$

4. $e^{-z} = 1/e^z$. In fact, by using Property 3, we see that $e^{-z}e^z = e^{-z+z} = 1$.
5. $e^z \neq 0$ for every $z \in \mathbf{C}$.
6. If $x \in \mathbf{R}$, then e^x, $\sin x$, and $\cos x$ are elements of $\mathbf{R}$, and in fact, all these functions when restricted to $\mathbf{R}$ coincide with the corresponding functions from real analysis.
7. The exponential function is a one-to-one increasing function on $\mathbf{R}$, and $\exp [\mathbf{R}] = \{x \colon x > 0\}$.
8. $e^{x+iy} = e^x e^{iy} = e^x(\cos y + i \sin y)$.
9. $|e^{x+iy}| = e^x$ and $\arg e^{x+iy} = y \bmod 2\pi$.
10. $|e^{x+iy}| = 1$ if and only if $x = 0$.
11. $e^z = e^w$ if and only if there exists $k \in \mathbf{Z}$ (where $\mathbf{Z}$ is the set of all integers) such that $z - w = 2\pi ki$. To see this, note that $e^z = e^w$ if and only if $e^{z-w} = 1$, and then apply Property 8.
12. In particular, $e^z = 1$ if and only if $z = 2\pi ki$ for some $k \in \mathbf{Z}$.
13. $\sin^2 z + \cos^2 z = 1$. Indeed,

$$\sin^2 z + \cos^2 z = \left(\frac{e^{iz} - e^{-iz}}{2i}\right)^2 + \left(\frac{e^{iz} + e^{-iz}}{2}\right)^2 = 1$$

14. $\cos(-z) = \cos z$ and $\sin(-z) = -\sin z$. That is, the cosine is an *even* function while the sine is an *odd* function.
15. Differentiating term by term yields $\sin' z = \cos z$ and $\cos' z = -\sin z$.
16. $\sin z = 0$ if and only if $z = k\pi$ for some $k \in \mathbf{Z}$, and $\cos z = 0$ if and only if $z = \pi/2 + k\pi$ for some $k \in \mathbf{Z}$. That is, the complex sine and cosine functions have the same zeros as the corresponding real functions.

We shall prove the result for the complex sine function. Observe that

$$0 = \sin z = \frac{e^{iz} - e^{-iz}}{2i} \quad \text{if and only if} \quad e^{iz} = e^{-iz}$$

which is equivalent to $e^{2iz} = 1$, and the result follows by Property 12.

A function $f \colon D \to \mathbf{C}$ is said to be *periodic* in a domain D if there exists a nonzero constant ω such that if $z \in D$, then $z + \omega \in D$ and $f(z + \omega) = f(z)$. Every such number ω is called a *period* of f. Clearly, if ω is a period of f in D, then for any nonzero integer n, $n\omega$ is a period of f. The set $\{n\omega \colon n \in \mathbf{Z}\}$ lies on the line L passing through 0 and ω; it constitutes a set of equidistant points. If $\omega_0 \in L$ is a period of f such that any other period $\omega \in L$ is of the form $\omega = n\omega_0$ for some nonzero integer n, then ω_0 is called a *fundamental* (or *primitive*) period of f. If f has no other periods in $\mathbf{C}$ except the nonzero integral multiples of a complex number ω, then f is called *simply periodic*.

Therefore Property 11 states that exp is a simply periodic function with primitive period $2\pi i$.

Example 3.6 Find the image of the strip $-\pi < \operatorname{Im} z \leq \pi$ under the mapping $w = e^z$.

Solution. Setting $z = x + iy$ and $w = u + iv$, we obtain $u = e^x \cos y$ and $v = e^x \sin y$, where $-\infty < x < \infty$ and $-\pi < y \leq \pi$. Observe that the line segment $x = 0$, $-\pi < y \leq \pi$ is mapped one-to-one and onto the circle $|w| = 1$. See Fig. 3.1.

The vertical line segment $x = x_0$, $-\pi < y \leq \pi$ is mapped one-to-one and onto the circle

$$u = e^{x_0} \cos y \qquad v = e^{x_0} \sin y \qquad -\pi < y \leq \pi$$

that is, the circle $|w| = e^{x_0}$. Since $e^{x_0} > 1$ for $x_0 > 0$ and $0 < e^{x_0} < 1$ for $x_0 < 0$, the right half of the strip is mapped onto the set $|w| > 1$, and the left half of the strip is mapped onto the set $0 < |w| < 1$. Thus the exponential function maps the strip $-\pi < \text{Im } z \leq \pi$ one-to-one and onto $\mathbf{C} - \{0\}$.

As in trigonometry, we define the other complex trigonometric functions, *tangent*, *cotangent*, *secant*, and *cosecant* by the formulas

$$\tan z = \frac{\sin z}{\cos z} \qquad \text{for } z \in \mathbf{C} - \left\{ \frac{\pi}{2} + n\pi : n \in \mathbf{Z} \right\}$$

$$\cot z = \frac{\cos z}{\sin z} \qquad \text{for } z \in \mathbf{C} - \{ n\pi : n \in \mathbf{Z} \}$$

$$\sec z = \frac{1}{\cos z} \qquad \text{for } z \in \mathbf{C} - \left\{ \frac{\pi}{2} + n\pi : n \in \mathbf{Z} \right\}$$

$$\csc z = \frac{1}{\sin z} \qquad \text{for } z \in \mathbf{C} - \{ n\pi : n \in \mathbf{Z} \}$$

Fig. 3.1

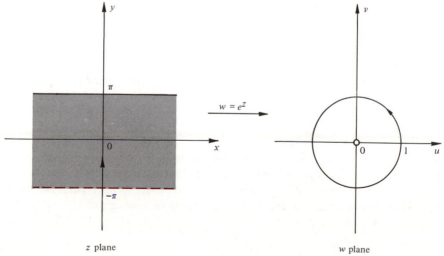

$w = e^z$

z plane

w plane

By a direct computation we see that the derivatives of the above functions are given by the following formulas (familiar from the calculus):

$$\tan' z = \sec^2 z \qquad \cot' z = -\csc^2 z$$

$$\sec' z = \sec z \tan z \qquad \csc' z = -\csc z \cot z$$

It is easily seen that all trigonometric functions are simply periodic. More precisely, we observe the following property.

17. The complex sine, cosine, secant, and cosecant functions have primitive periods equal to 2π, whereas the complex tangent and cotangent functions have primitive periods equal to π.

Clearly, all the complex trigonometric functions are analytic. Using Euler's formulas, one can easily show that every algebraic identity involving real trigonometric functions yields a valid identity for the complex trigonometric functions.

In Sec. 4.5 we shall show that if D is a domain, if f and g are analytic functions on D, and if $S \subset D$ is a set with a limit point in D such that $f(z) = g(z)$ for all $z \in S$, then $f(z) = g(z)$ for all $z \in D$. An immediate consequence of this statement is that every algebraic identity involving real trigonometric functions yields a valid identity for the complex trigonometric functions. For example, in order to see that

$$\cos (z + w) = \cos z \cos w - \sin z \sin w \qquad \text{for all } z, w \in \mathbf{C}$$

it suffices to recall that the above identity is valid if z and w are real numbers. Finally the hyperbolic functions are defined by

$$\cosh z = \frac{e^z + e^{-z}}{2} \qquad \sinh z = \frac{e^z - e^{-z}}{2} \qquad \text{for } z \in \mathbf{C}$$

$$\tanh z = \frac{\sinh z}{\cosh z} \qquad \text{for } z \in \mathbf{C} - \{(n + {}^1\!/_2)\pi i : n \in \mathbf{Z}\}$$

$$\coth z = \frac{\cosh z}{\sinh z} \qquad \text{for } z \in \mathbf{C} - \{n\pi i : n \in \mathbf{Z}\}$$

$$\operatorname{sech} z = \frac{1}{\cosh z} \qquad \text{for } z \in \mathbf{C} - \{(n + {}^1\!/_2)\pi i : n \in \mathbf{Z}\}$$

$$\operatorname{csch} z = \frac{1}{\sinh z} \qquad \text{for } z \in \mathbf{C} - \{n\pi i : n \in \mathbf{Z}\}$$

The properties of each of these functions again follow from the properties of the exponential function. For example, $\cosh z$ and $\sinh z$ are entire functions, $\cosh' z = \sinh z$ and $\sinh' z = \cosh z$, and they are simply periodic with primitive period equal to $2\pi i$. The only zeros of $\cosh z$ are

$$\{(n + {}^1\!/_2)\pi i : n \in \mathbf{Z}\}$$

and the only zeros of sinh z are $\{n\pi i : n \in \mathbf{Z}\}$. The functions tanh z, coth z, sech z, and csch z are analytic functions in their domain of definition, and they satisfy

$$\text{tanh}' z = \text{sech}^2 z \qquad\qquad \text{coth}' z = -\text{csch}^2 z$$

$$\text{sech}' z = -\text{sech } z \text{ tanh } z \qquad \text{csch}' z = -\text{csch } z \text{ coth } z$$

Continuing our list of properties, we can easily verify that

18. $\cosh z = \cos iz$.
19. $\sinh z = -i \sin iz$.
20. $\sin (x + iy) = \sin x \cosh y + i \cos x \sinh y$.
21. $\cos (x + iy) = \cos x \cosh y - i \sin x \sinh y$.

Example 3.7 Consider the mapping $w = \sin z$ of the z plane into the w plane. (*a*) Find the image of every horizontal and vertical line in the z plane under this mapping. (*b*) Show that the semi-infinite strip

$$|\text{Re } z| < \frac{\pi}{2} \qquad \text{Im } z > 0$$

is mapped one-to-one and onto the upper half-plane Im $w > 0$. (*c*) Show that the boundary of the semi-infinite strip in (*b*) is mapped onto the real axis in the w plane.

Solution
a. We have

$$u + iv = w = \sin (x + iy) = \sin x \cosh y + i \cos x \sinh y$$

Hence

$$u = \sin x \cosh y \qquad v = \cos x \sinh y \qquad\qquad (3.3)$$

Let us start with the horizontal lines first. Clearly the line $y = 0$ is mapped onto the segment $u = \sin x$, $v = 0$, $-\infty < x < \infty$. That is, $v = 0$, $-1 \le u \le 1$. This segment $-1 \le u \le 1$, $v = 0$ is traversed infinitely often when the point z describes the entire straight line $y = 0$. The segment $-\pi/2 \le x \le \pi/2$, $y = 0$ on the line $y = 0$ is mapped one-to-one and onto the segment $-1 \le u \le 1$, $v = 0$. If $y = y_0 \ne 0$, we obtain from (3.3)

$$\frac{u^2}{\cosh^2 y_0} + \frac{v^2}{\sinh^2 y_0} = 1 \qquad\qquad (3.4)$$

This is the equation of an ellipse with half-axes $\cosh y_0$ and $\sinh y_0$. Since $\cosh^2 y_0 - \sinh^2 y_0 = 1$, the foci of the ellipse are at ± 1. As z describes the line segment $y = y_0$, its image w describes the ellipse (3.4) infinitely often. This property can be seen from (3.3), which also proves that any segment $x_0 \le x < x_0 + 2\pi$, $y = y_0$ of the line $y = y_0$ is mapped one-to-one and onto the ellipse (3.4). The image of the line $y = 0$

can be regarded as a degenerate ellipse. Next, we will find the image of any vertical line. Clearly if $n \in \mathbf{Z}$, then the line $x = n\pi$ is mapped onto

$$u = (-1)^n \cosh y \qquad v = 0 \qquad -\infty < y < \infty$$

that is, it is mapped onto the segment $-\infty < u \le -1$, $v = 0$ when n is odd and onto the segment $1 \le u < \infty$, $v = 0$ when n is even. If $x = x_0$ with x_0 not equal to $n\pi$ or $n\pi + (\pi/2)$ for any $n \in \mathbf{Z}$, we obtain from Eq. (3.3)

$$\frac{u^2}{\sin^2 x_0} - \frac{v^2}{\cos^2 x_0} = 1 \tag{3.5}$$

This is the equation of a hyperbola with half-axes $\sin x_0$ and $\cos x_0$. Since $\sin^2 x_0 + \cos^2 x_0 = 1$, the foci of this hyperbola are at ± 1. The images of $x_0 = n\pi$ and $x_0 = n\pi + (\pi/2)$ can be regarded as degenerate hyperbolas.

b. From the discussion in part (a) of this solution, we see that the images of the segments $|x| < \pi/2$, $y = y_0 > 0$ are the upper halves of the ellipses (3.4). As y_0 varies in $(0,\infty)$, these half-ellipses clearly sweep out the half-plane Im $w > 0$. See Fig. 3.2.

c. Again from the discussion in part (a), we see that the image of the half-line $x = -\pi/2$, $y \ge 0$ is the half-line $-\infty < u \le -1$, $v = 0$; the image of the segment $-\pi/2 \le x \le \pi/2$, $y = 0$ is the segment $-1 \le u \le 1$, $v = 0$; and the image of the half-line $x = \pi/2$, $y \ge 0$ is the half-line $1 \le u < \infty$, $v = 0$.

EXERCISES

3.4.1 Show that (a) $e^{\pi i} = -1$; (b) $e^{2\pi i} = 1$; (c) $e^{1 + (\pi/2)i} = ie$.

3.4.2 Show the following.

$$a.\ \lim_{z \to 0} \frac{e^z - 1}{z} = 1 \qquad b.\ \lim_{z \to 0} \frac{\sin z}{z} = 1 \qquad c.\ \lim_{z \to 0} \frac{\cos z - 1}{z} = 0$$

Fig. 3.2

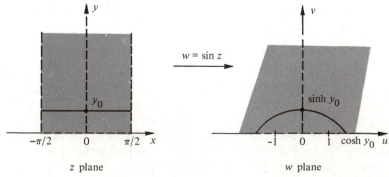

$w = \sin z$

z plane

w plane

3.4.3 Solve the equations.

    a. $e^z = -1 + i$ *b.* $e^z = i$ *c.* $\sin z = 100$

    d. $\tan z = i$ (Can you now resolve the apparent paradox on page 2?)

    e. $\cos^2 z = -9$ *f.* $\sin z + \cos z = i$

3.4.4 Show that, for $z = x + iy$, the following hold.

    a. $|\sin z|^2 = \sin^2 x + \sinh^2 y = -\cos^2 x + \cosh^2 y$

    b. $|\sin x| \le |\sin z|$

    c. $|\cos z|^2 = \cos^2 x + \sinh^2 y = -\sin^2 x + \cosh^2 y$

    d. $|\cos x| \le |\cos z|$

    e. $|\sin z| \le \cosh y$ and $|\sin z| \ge |\sinh y|$

    f. $|\cos z| \le \cosh y$ and $|\cos z| \ge |\sinh y|$

3.4.5 Show that for $z = x + iy \ne \pi/2 + n\pi$, $n \in \mathbf{Z}$,

$$\tan z = \frac{\sin 2x + i \sinh 2y}{\cos 2x + \cosh 2y}$$

3.4.6 Does the equation $\tan z = z$ have any nonreal root? How about the equation $z \tan z = 1$?

3.4.7 Show that the mapping $w = \sin z$ maps the semi-infinite strip $0 < \text{Re } z < \pi/2$, $\text{Im } z > 0$ one-to-one and onto the first quadrant $\text{Re } w > 0$, $\text{Im } w > 0$.

3.4.8 Find the image of the semi-infinite strip $|\text{Re } z| < \pi/2$, $\text{Im } z > 0$ under the mapping $w = \cos z$.

3.4.9 Show that

    a. $\cos(\pi/2 - z) = \sin z$ *b.* $\sin(\pi/2 - z) = \cos z$

    c. $\cos(\pi - z) = -\cos z$ *d.* $\sin(\pi - z) = \sin z$

    e. $\tan(\pi + z) = \tan z$ *f.* $\cot(\pi/2 - z) = \tan z$

3.4.10 Show the following.

    a. $\cosh^2 z - \sinh^2 z = 1$

    b. $\cosh 2z = \cosh^2 z + \sinh^2 z$

    c. $\sinh 2z = 2 \sinh z \cosh z$

    d. $\sinh(i\pi/2 - z) = i \cosh z$

3.4.11 Find all the zeros and periods of $\cosh z$, $\sinh z$, $\tanh z$, and $\coth z$.

3.4.12 Show that if $0 < |z| < 1$, then

    a. $\dfrac{|z|}{4} < |e^z - 1| < \tfrac{7}{4}|z|$

    b. $|\cos z| < 2$

    c. $|\sin z| < \tfrac{13}{10}|z|$

3.5

THE COMPLEX LOGARITHM

    In this section, we define a complex analog of the logarithm function of real analysis; that is, given a complex number z, we wish to determine all complex numbers w such that $e^w = z$. As $e^w \ne 0$ for all $w \in \mathbf{C}$, we see that we must exclude 0 from the domain of the logarithm. Hence if $z \in \mathbf{C} - \{0\}$, we define

$$\log z = \{w: e^w = z\}$$

Recall that if we write $z = r(\cos \theta + i \sin \theta)$ where $\theta \in \arg z$, then we have

$$z = |z|e^{i\theta} = e^{\ln |z| + i \arg z}$$

where $\ln |z|$ denotes the natural logarithm of the positive number $|z|$. Thus

$$\log z = \ln |z| + i \arg z$$

The value

$$\text{Log } z = \ln |z| + i \text{ Arg } z \qquad -\pi < \text{Arg } z \leq \pi$$

is called the *principal value* of the logarithm of z. Clearly if z is a positive real number, then Log $z = \ln z$. Also, if $z \in \mathbf{C} - \{0\}$, then

$$\text{Log } z = \text{Log } |z| + i \text{ Arg } z$$

The following properties of the logarithm are easily established for nonzero complex numbers.

1. $e^{\log z} = z$ (more precisely, $e^{\log z} = \{z\}$)
2. $\log e^z = z \mod 2\pi i$
3. $\log z_1 z_2 = \log z_1 + \log z_2 \mod 2\pi i$
4. $\log (z_1/z_2) = \log z_1 - \log z_2 \mod 2\pi i$
5. $\log z^n = n \log z \mod 2\pi i$ for all $n \in \mathbf{Z}$

If D is a domain in $\mathbf{C}$ and $f: D \to \mathbf{C}$ is a continuous function satisfying $e^{f(z)} = z$ for all $z \in D$, then we call f a *branch* of log.

We are now ready to show that while Log is not even continuous on the negative real axis, if we restrict Log to the domain $\{z \in \mathbf{C}: z + |z| \neq 0\}$ (that is, the whole complex plane minus the nonpositive real axis), we obtain a branch of log called the *principal branch*. This results from the following theorem.

Theorem 3.3 The function Arg is (*a*) continuous on $\{z \in \mathbf{C}: z + |z| \neq 0\}$, (*b*) nowhere continuous on $\{z \in \mathbf{C}: z + |z| = 0\}$.

Proof
a. Let $D = \{z \in \mathbf{C}: z + |z| \neq 0\}$. It suffices to show that if $z_1, z_2, \ldots$ is a sequence in D, $z \in D$, and $\lim_{n\to\infty} z_n = z$, then $\lim_{n\to\infty} \text{Arg } z_n = \text{Arg } z$. In fact, from the identity

$$\frac{z_n}{|z_n|} = e^{i \text{ Arg } z_n} \qquad n = 1, 2, \ldots$$

we obtain

$$e^{i \text{ Arg } z} = \frac{z}{|z|} = \lim_{n\to\infty} \frac{z_n}{|z_n|} = \lim_{n\to\infty} e^{i \text{ Arg } z_n} \qquad (3.6)$$

We must next show that $\lim_{n \to \infty} \operatorname{Arg} z_n$ exists. If it does not, then since $\{\operatorname{Arg} z_n : n = 1, 2, \ldots\}$ is a bounded set of real numbers,

$$s = \inf \{\operatorname{Arg} z_n : n = 1, 2, \ldots\} \qquad \text{and} \qquad S = \sup \{\operatorname{Arg} z_n : n = 1, 2, \ldots\}$$

both exist, and in fact both are numbers in $[-\pi, \pi]$. Thus there exist subsequences $\{z_{k(n)}\}$ and $\{z_{l(n)}\}$ such that $\lim_{n \to \infty} \operatorname{Arg} z_{k(n)} = s$ and $\lim_{n \to \infty} \operatorname{Arg} z_{l(n)} = S$. Hence by Eq. (3.6) and the continuity of the exponential function,

$$e^{is} = \lim_{n \to \infty} e^{i \operatorname{Arg} z_{k(n)}} = e^{i \operatorname{Arg} z} = \lim_{n \to \infty} e^{i \operatorname{Arg} z_{l(n)}} = e^{iS}$$

So $S - s = 2\pi k$ for some $k \in \mathbf{Z}$. In view of the fact that $-\pi < \operatorname{Arg} \zeta \le \pi$ for all $\zeta \in D$, we see that $k = 0$ or $k = 1$. If $k = 1$, then $s = -\pi$ and $S = \pi$, which implies that $z/|z| = e^{i\pi} = -1$. This is impossible because $z \in D$. Hence $k = 0$ and so $\lim_{n \to \infty} \operatorname{Arg} z_n$ does exist. So again by Eq. (3.6) and the continuity of the exponential function,

$$e^{i \operatorname{Arg} z} = \lim_{n \to \infty} e^{i \operatorname{Arg} z_n} = e^{i \lim_{n \to \infty} \operatorname{Arg} z_n}$$

and so as before, we have $\operatorname{Arg} z = \lim_{n \to \infty} \operatorname{Arg} z_n$, which concludes the proof of part (a) of this theorem.

b. Let $x \le 0$. We shall show that Arg is discontinuous at x. Let $z_n = x + (i/n)$ and let $\zeta_n = x - (i/n)$ for $n = 1, 2, \ldots$. Then $\lim_{n \to \infty} z_n = x = \lim_{n \to \infty} \zeta_n$, while $\lim_{n \to \infty} \operatorname{Arg} z_n = \pi$ and $\lim_{n \to \infty} \operatorname{Arg} \zeta_n = -\pi$, which concludes the proof.

Note that in part (b) of Theorem 3.3 we actually proved that if $g(z) = \operatorname{Arg} z$ for $z \in \{\zeta \in \mathbf{C} : \zeta + |\zeta| \ne 0\}$, then it is impossible to extend the definition of g to make it continuous at any point of the nonpositive real axis. Hence it is impossible to extend the principal branch of log to be continuous at any point of $\{z \in \mathbf{C} : z + |z| = 0\}$.

Theorem 3.4 Let $f : D \to \mathbf{C}$ be a branch of log. Then a function $g : D \to \mathbf{C}$ is a branch of log if and only if there exists an integer k such that $g(z) = f(z) + 2k\pi i$ for all $z \in D$.

Proof. If $k \in \mathbf{Z}$ and $g(z) = f(z) + 2k\pi i$ for all $z \in D$, clearly g is a branch of log. So let g be a branch of log. It suffices for us to show the existence of an integer k satisfying $g(z) = f(z) + 2k\pi i$ for all $z \in D$. Let $h(z) = g(z) - f(z)$ if $z \in D$. Then h is continuous as g and f are, and if $z \in D$,

$$e^{h(z)} = e^{g(z) - f(z)} = \frac{e^{g(z)}}{e^{f(z)}} = \frac{z}{z} = 1$$

Hence the composite function $\exp \circ h$ is constantly equal to one, and so $h : D \to \{2\pi k i : k \in \mathbf{Z}\}$. So as D is arcwise connected and $\{2\pi k i : k \in \mathbf{Z}\}$ is discrete, it follows by Exercise 2.6.1 that h is constant, which proves the theorem.

Theorem 3.5 Let $f: D \to \mathbf{C}$ be a branch of log. Then $f'(z) = 1/z$ for all $z \in D$ and so in particular, f is analytic.

Proof. Note that f is one-to-one since $f(z) = f(w)$ implies $z = e^{f(z)} = e^{f(w)} = w$. Hence

$$f'(z) = \lim_{h \to 0} \frac{f(z + h) - f(z)}{h} = \lim_{h \to 0} \frac{f(z + h) - f(z)}{e^{f(z+h)} - e^{f(z)}}$$

$$= \frac{1}{\displaystyle\lim_{h \to 0} \frac{e^{f(z+h)} - e^{f(z)}}{f(z + h) - f(z)}} \qquad \text{(as } f \text{ is one-to-one)}$$

$$= \frac{1}{\exp' f(z)} \qquad \text{(as } f \text{ is continuous at } z\text{)}$$

$$= \frac{1}{\exp f(z)} = \frac{1}{z}$$

Example 3.8 Show that (a) $\text{Log}\,(-1) = i\pi$; (b) $\log i = i(\pi/2)$ mod $2\pi i$; (c) $\text{Log}\,[i(-1 + i)] \neq \text{Log}\,i + \text{Log}\,(-1 + i)$; (d) $\text{Log}\,(-1)^2 \neq 2\,\text{Log}\,(-1)$.

Proof
a. $|-1| = 1$ and $\text{Arg}\,(-1) = \pi$, hence $\text{Log}\,(-1) = \text{Log}\,1 + i\pi = i\pi$.
b. $|i| = 1$ and $\text{Arg}\,i = \pi/2$, hence $\log i = \text{Log}\,1 + i(\pi/2)$ mod $2\pi i$.
c. $i(-1 + i) = -1 - i$ with $|-1 - i| = \sqrt{2}$ and $\text{Arg}\,(-1 - i) = -3\pi/4$. Hence

$$\text{Log}\,[i(-1 + i)] = \text{Log}\,\sqrt{2} - i\frac{3\pi}{4}$$

On the other hand, $|-1 + i| = \sqrt{2}$, $\text{Arg}\,(-1 + i) = 3\pi/4$, and hence $\text{Log}\,(-1 + i) = \text{Log}\,\sqrt{2} + i(3\pi/4)$. Also $\text{Log}\,i = i(\pi/2)$. Comparing these values, (c) follows.
d. $\text{Log}\,(-1)^2 = \text{Log}\,1 = 0$, but from part (a) we can see that $2\,\text{Log}\,(-1) = 2\pi i \neq 0$.

Now we define what is meant by a complex number raised to a complex number. If z and a are complex numbers with $z \neq 0$, we define

$$z^a = e^{a \log z}$$

That is, $z^a = \{e^{aw} : w \in \log z\}$.
The *principal value* of z^a is by definition $e^{a\,\text{Log}\,z}$.
In order not to conflict with our previous terminology, we agree that unless otherwise specified, e^a stands for its principal value.

Example 3.9 Compute i^i and 1^i. Find also their principal values.

Solution. We have

$$i^i = e^{i \log i} = \{e^{i[i(\pi/2) + 2\pi i k]}: k \in \mathbf{Z}\}$$

$$= \{e^{(-\pi/2) - 2\pi k}: k \in \mathbf{Z}\}$$

The principal value of i^i is $e^{-\pi/2}$. Also,

$$1^i = e^{i \log 1} = \{e^{i(2\pi i k)}: k \in \mathbf{Z}\} = \{e^{-2\pi k}: k \in \mathbf{Z}\}$$

and the principal value of 1^i is 1.

In general z^a has infinitely many distinct elements. However if a is rational, say, $a = m/n$ where n is a positive integer and m and n are relatively prime, then $z^{m/n}$ has only n distinct values. In fact,

$$z^{m/n} = e^{(m/n) \log z} = e^{(m/n) \operatorname{Log} z} e^{(m/n)2\pi i k} \qquad k = 0, \pm 1, \pm 2, \ldots$$

Clearly, for $k = 0, 1, 2, \ldots, n - 1$ we get n distinct values, but for $k = n + l$ we get the same value as when $k = l$. In fact,

$$e^{(m/n)2\pi i(n+l)} = e^{2\pi i m} e^{(m/n)2\pi i l} = e^{(m/n)2\pi i l}$$

This result agrees with the result of Sec. 1.5 when m and n are relatively prime.

The addition theorem $z^{a_1} z^{a_2} = z^{a_1 + a_2}$ holds for the principal values of these powers. In fact,

$$z^{a_1} z^{a_2} = e^{a_1 \operatorname{Log} z} e^{a_2 \operatorname{Log} z}$$

$$= e^{(a_1 + a_2) \operatorname{Log} z} = z^{a_1 + a_2}$$

However, for the principal values, in general

$$(z_1 z_2)^a \neq z_1^a z_2^a$$

$$\left(\frac{z_1}{z_2}\right)^a \neq \frac{z_1^a}{z_2^a}$$

$$\operatorname{Log} z^a \neq a \operatorname{Log} z$$

$$(z^a)^b \neq z^{ab}$$

where by $(z^a)^b$ we mean $\exp (b \operatorname{Log} e^{a \operatorname{Log} z})$.

Example 3.10 Let $D = \mathbf{C} - \{x \in \mathbf{R}: x \leq 0\}$, let $a \in \mathbf{C}$, and let f and g be functions from D to $\mathbf{C}$ defined by

$$f(z) = e^{a \operatorname{Log} z} \qquad \text{and} \qquad g(z) = ae^{(a-1) \operatorname{Log} z}$$

Then f is analytic on D, and if $z \in D$, then

$$f'(z) = g(z)$$

This is often written as

$$(z^a)' = az^{a-1}$$

where z^a and z^{a-1} are taken to be the principal values of the corresponding power functions.

Proof. As

$$f(z) = \exp{(a \operatorname{Log} z)} \qquad \text{for } z \in D$$

$$f'(z) = \exp'{(a \operatorname{Log} z)}\, a \operatorname{Log}' z$$

$$= e^{a \operatorname{Log} z} a\, \frac{1}{z} = a e^{a \operatorname{Log} z} e^{-\operatorname{Log} z}$$

$$= a e^{(a-1)\operatorname{Log} z} = g(z)$$

Finally, we shall indicate to the reader how the *complex inverse trigonometric functions* are defined. We shall consider $\cos^{-1}$ here. The others can be defined similarly and are left to the reader in the exercises. Let $z \in \mathbf{C}$ and $w = \cos z$. We wish to define $\cos^{-1} w$. Recall that

$$w = \cos z = \frac{e^{iz} + e^{-iz}}{2}$$

Hence $e^{2iz} - 2we^{iz} + 1 = 0$, and so $e^{iz} = w + \sqrt{w^2 - 1}$; here as before, we suppress the $\pm$ sign in front of $\sqrt{w^2 - 1}$, as $\sqrt{w^2 - 1}$ has two values. So we have $iz = \log (w + \sqrt{w^2 - 1})$ or $z = -i \log (w + \sqrt{w^2 - 1})$. Hence we define $\cos^{-1} w = -i \log (w + \sqrt{w^2 - 1})$ for every $w \in \mathbf{C}$. In other words, $\cos^{-1} w = \{-i\zeta : \zeta \in \log (w + z) \text{ for some } z \in \sqrt{w^2 - 1}\}$.

EXERCISES

3.5.1 Compute the following:

a. $\log (1 + i)$

b. $\operatorname{Log} (1 + i)^2$

c. $\log e$

d. $\operatorname{Log} e$

e. $\operatorname{Log} (1 + i)^i$

f. $(\sqrt{3} - i)^i$

g. 2^i

h. $16^{1/4}$

i. $\operatorname{Re} (1 + i)^{1+i}$

j. $(1 - i)^{1/4}$

k. Find the principal value of $(-1 + i)^{-i}$.

l. $i^{1/5}$

m. $|(-1)^i|$

n. $(2 - i)^i (2 - i)^{-i}$ for the principal values of the powers.

3.5.2 Show that

a. $\sin^{-1} z = -i \log (iz + \sqrt{z^2 - 1})$

b. $\tan^{-1} z = \frac{i}{2} \log \frac{i + z}{i - z}$

3.5.3 Compute $\cosh^{-1} z$, $\sinh^{-1} z$, and $\tanh^{-1} z$.

3.6

THE CAUCHY-RIEMANN EQUATIONS

Up to this point we have stressed the similarities of the theory of complex functions with that of real functions. However, we have now reached a point of departure. If $z = x + iy$ and

$$f(z) = u(x,y) + iv(x,y)$$

the existence of partial derivatives of u and v does not imply the differentiability of f without the expense of additional hypotheses. A trivial example to see this is given by the function $f(z) = \bar{z}$. In this case,

$$u(x,y) = x \qquad \text{and} \qquad v(x,y) = -y$$

both have continuous partial derivatives, but as we proved in Example 3.2, the function f is nowhere differentiable in the complex plane $\mathbf{C}$. In this section we give necessary and sufficient conditions for a function f to have a derivative at a point z_0 in its domain of definition. The results that we shall prove give, in turn, necessary and sufficient conditions for a function f to be analytic in a domain D. The following result gives necessary conditions for a function $f: D \to \mathbf{C}$ to be differentiable at a point $z_0 \in D$.

Theorem 3.6 Assume that $f(z) = u(x,y) + iv(x,y)$ is differentiable at a point $z_0 = x_0 + iy_0 \in D$. Then the four partial derivatives $u_x(x_0,y_0)$, $u_y(x_0,y_0)$, $v_x(x_0,y_0)$, and $v_y(x_0,y_0)$ exist and satisfy the *Cauchy-Riemann equations*

$$\begin{align} u_x(x_0,y_0) &= v_y(x_0,y_0) \\ u_y(x_0,y_0) &= -v_x(x_0,y_0) \end{align} \tag{3.7}$$

Proof. Since, by assumption, the derivative $f'(z_0)$ exists, the difference quotient

$$\frac{f(z_0 + h) - f(z_0)}{h}$$

must approach the limit $f'(z_0)$ as h approaches zero in any manner. We first let $h \to 0$ through real values. Then

$$f'(z_0) = \lim_{h \to 0} \frac{f(z_0 + h) - f(z_0)}{h}$$

$$= \lim_{h \to 0} \frac{u(x_0 + h, y_0) - u(x_0,y_0)}{h} + i \lim_{h \to 0} \frac{v(x_0 + h, y_0) - v(x_0,y_0)}{h}$$

This proves (see Exercise 2.3.5) that the partial derivatives $u_x(x_0,y_0)$ and $v_x(x_0,y_0)$ exist, and moreover that

$$f'(z_0) = u_x(x_0,y_0) + iv_x(x_0,y_0) \tag{3.8}$$

Next, let $h = it \to 0$ (t real) through imaginary values. Then

$$f'(z_0) = \lim_{t \to 0} \frac{f(z_0 + it) - f(z_0)}{it}$$

$$= -i \lim_{t \to 0} \frac{u(x_0, y_0 + t) - u(x_0, y_0)}{t} + \lim_{t \to 0} \frac{v(x_0, y_0 + t) - v(x_0, y_0)}{t}$$

GEORG FRIEDRICH BERNHARD RIEMANN

Georg Friedrich Bernhard Riemann was born at Breselenz-Hanover, Germany, on September 17, 1826. His father was a Lutheran pastor and his mother was the daughter of a court counselor.

Riemann was brought up in a warm family atmosphere, although he was shy and of poor health (the latter due to poverty at home). Thanks to his father's understanding, Riemann did not practice theology, for which he was trained at the University of Göttingen and Berlin. He obtained his doctorate from Göttingen in 1851 with a celebrated dissertation on the theory of complex functions. It is here that we find the so-called Cauchy-Riemann equations [see Eq. (3.7)]. Riemann is considered one of the three founders of complex-function theory (the others being Cauchy and Weierstrass), and he contributed significantly to many other areas of mathematics and physics. In fact, he initiated the study of many more topics than he could ever finish. In 1859 Riemann succeeded Dirichlet as professor of mathematics at Göttingen.

At the age of 36, almost three years after his appointment as full professor at Göttingen, Riemann married his sister's friend Elise Koch. One daughter was born of this marriage, in August 1863. Unfortunately, several days after his marriage he fell ill with pleurisy. With the financial support of the German government, he visited Italy hoping that its mild climate would improve his condition, which it did for a short while. He made two subsequent trips to Italy for the same reason, but he never returned from his last one. Riemann died in a villa at Selasca, Lago Maggiere, on July 20, 1866. He was only 39 years old, too young to have completed all the things that he had started.

This proves that the partial derivatives $u_y(x_0,y_0)$ and $v_y(x_0,y_0)$ exist and

$$f'(z_0) = -iu_y(x_0,y_0) + v_y(x_0,y_0) \qquad (3.9)$$

Hence the four partial derivatives u_x, u_y, v_x, and v_y exist at (x_0,y_0), and from Eqs. (3.8) and (3.9) we see that the Cauchy-Riemann equations (3.7) are satisfied. The proof is complete.

The following example, due to D. Menchoff in 1936, shows that the converse of Theorem 3.6 is not, in general, true.

Example 3.11 Consider the function $f: \mathbf{C} \to \mathbf{C}$ defined by

$$f(z) = \begin{cases} \dfrac{z^5}{|z|^4} & \text{if } z \neq 0 \\ 0 & \text{if } z = 0 \end{cases}$$

Show that u_x, u_y, v_x, and v_y exist and satisfy the Cauchy-Riemann equations at $(0,0)$, but that $f'(0)$ does not exist.

Proof. Setting $z = x + iy$, we obtain

$$u(x,y) = \begin{cases} \dfrac{x^5 - 10x^3y^2 + 5xy^4}{(x^2 + y^2)^2} & (x,y) \neq (0,0) \\ 0 & (x,y) = (0,0) \end{cases}$$

and

$$v(x,y) = \begin{cases} \dfrac{5x^4y - 10x^2y^3 + y^5}{(x^2 + y^2)^2} & (x,y) \neq (0,0) \\ 0 & (x,y) = (0,0) \end{cases}$$

From the definition of partial derivatives for real-valued functions we obtain

$$u_x(0,0) = \lim_{h \to 0} \frac{u(h,0) - u(0,0)}{h} = \lim_{h \to 0} \frac{h^5}{h^5} = 1$$

and

$$u_y(0,0) = \lim_{h \to 0} \frac{u(0,h) - u(0,0)}{h} = \lim_{h \to 0} \frac{0}{h^5} = 0$$

Similarly, we obtain $v_x(0,0) = 0$ and $v_y(0,0) = 1$. Hence, $u_x(0,0)$, $u_y(0,0)$, $v_x(0,0)$, and $v_y(0,0)$ exist and satisfy the Cauchy-Riemann equations (3.7). However $f'(0)$ does not exist, since

$$\frac{f(h)}{h} = \left(\frac{h}{|h|} \right)^4$$

has the value 1 when $h > 0$ and takes the value -1 when $h = h_1 + ih_2$ with $h_1 = h_2 > 0$.

A consequence of Theorem 3.6 is the following corollary.

Corollary 3.4 Let $f: D \to \mathbf{C}$, and for $z = x + iy \in D$ let

$$f(z) = u(x,y) + iv(x,y)$$

A necessary condition for f to be analytic in the domain D is that the four partial derivatives u_x, u_y, v_x, and v_y exist and satisfy the Cauchy-Riemann equations

$$u_x = v_y \qquad u_y = -v_x$$

at each point of D.

Example 3.12 The function $f(z) = \bar{z}$ is nowhere analytic in $\mathbf{C}$.

Proof. In fact, in this case

$$\begin{aligned}
u(x,y) &= x & v(x,y) &= -y \\
u_x(x,y) &= 1 & u_y(x,y) &= 0 \\
v_x(x,y) &= 0 & v_y(x,y) &= -1
\end{aligned}$$

and the Cauchy-Riemann equations are not satisfied at any point of $\mathbf{C}$.

The following result gives sufficient conditions for a function $f: D \to \mathbf{C}$ to be differentiable at a point $z_0 \in D$.

Theorem 3.7 Let $f(z) = u(x,y) + iv(x,y)$ and $z_0 = x_0 + iy_0 \in D$. Assume that the four partial derivatives u_x, u_y, v_x, and v_y exist and are continuous in an ε neighborhood of z_0 and in addition satisfy the Cauchy-Riemann equations

$$u_x(x_0,y_0) = v_y(x_0,y_0) \qquad u_y(x_0,y_0) = -v_x(x_0,y_0) \tag{3.7}$$

at (x_0,y_0). Then f is differentiable at z_0.

Proof. Let $h = h_1 + ih_2$ be any complex number such that $z_0 + h \in N_\varepsilon(z_0)$. Since u and v have continuous first partial derivatives in $N_\varepsilon(z_0)$, it follows from the mean value theorem of the calculus that

$$\frac{f(z_0 + h) - f(z_0)}{h}$$

$$= \frac{u(x_0 + h_1, y_0 + h_2) - u(x_0,y_0)}{h}$$

$$+ i\frac{v(x_0 + h_1, y_0 + h_2) - v(x_0,y_0)}{h}$$

$$= \frac{u_x(x_0 + \theta_1 h_1, y_0 + \theta_1 h_2)h_1 + u_y(x_0 + \theta_1 h_1, y_0 + \theta_1 h_2)h_2}{h}$$

$$+ i\frac{v_x(x_0 + \theta_2 h_1, y_0 + \theta_2 h_2)h_1 + v_y(x_0 + \theta_2 h_1, y_0 + \theta_2 h_2)h_2}{h}$$

$$= \frac{u_x(x_0,y_0)h_1 + u_y(x_0,y_0)h_2}{h} + i\frac{v_x(x_0,y_0)h_1 + v_y(x_0,y_0)h_2}{h}$$

$$+ \varepsilon_1(h)\frac{h_1}{h} + \varepsilon_2(h)\frac{h_2}{h}$$

where $0 \le \theta_1, \theta_2 \le 1$, and $\lim_{h \to 0} \varepsilon_1(h) = 0 = \lim_{h \to 0} \varepsilon_2(h)$. In view of the Cauchy-Riemann equations and the fact that

$$\left| \frac{h_j}{h} \right| \le 1 \qquad j = 1, 2$$

we obtain

$$\frac{f(z_0 + h) - f(z_0)}{h} = u_x(x_0, y_0) + iv_x(x_0, y_0) + \varepsilon_3(h)$$

where $\lim_{h \to 0} \varepsilon_3(h) = 0$. Taking limits as $h \to 0$, we see that $f'(z_0)$ exists and is equal to $u_x(x_0, y_0) + iv_x(x_0, y_0)$. The proof is complete.

A consequence of Theorem 3.7 is the following.

Corollary 3.5 Let $f: D \to \mathbf{C}$, and for $z = x + iy$ let

$$f(z) = u(x, y) + iv(x, y)$$

A sufficient condition for f to be analytic in D is that the four partial derivatives u_x, u_y, v_x, and v_y exist, be continuous, and satisfy the Cauchy-Riemann equations

$$u_x = v_y \qquad u_y = -v_x$$

at each point of D.

Example 3.13 Using the methods of this section, show that $f(z) = z^2 + 3z + i$ is an entire function.

Proof. In fact, in this case

$$
\begin{aligned}
u(x,y) &= x^2 - y^2 + 3x & v(x,y) &= 2xy + 3y + 1 \\
u_x &= 2x + 3 & v_x &= 2y \\
u_y &= -2y & v_y &= 2x + 3
\end{aligned}
$$

and all the hypotheses of Corollary 3.5 are satisfied at every point of $\mathbf{C}$.

The sufficiency condition of Corollary 3.5 includes the requirement that the partial derivatives u_x, u_y, v_x, and v_y be continuous. In Theorem 4.5 we shall prove the converse of this; that is, if f is analytic, then f' is also analytic, and therefore u_x, u_y, v_x, and v_y are continuous. Using this observation we can state the following theorem.

Theorem 3.8 A necessary and sufficient condition for a function $f(z) = u(x,y) + iv(x,y)$ to be analytic in a domain D is that the four partial derivatives u_x, u_y, v_x, and v_y exist, be continuous, and satisfy the Cauchy-Riemann equations

$$u_x = v_y \qquad u_y = -v_x$$

at each point of D.

3.6.1 Using the Cauchy-Riemann equations, show that the functions $e^{\bar{z}}$, $\sin \bar{z}$, and $\cos \bar{z}$ are nowhere analytic.

3.6.2 Using the Cauchy-Riemann equations, show that the function

$$f(z) = (e^x \cos y + 3) + i(e^x \sin y + 5) \qquad \text{for } z = x + iy$$

is analytic everywhere in **C**.

3.6.3 Show that the functions

$$f_1(z) = \begin{cases} \dfrac{\bar{z}^3}{|z|^2} & z \neq 0 \\ 0 & z = 0 \end{cases}$$

and

$$f_2(z) = \begin{cases} e^{-1/z^4} & z \neq 0 \\ 0 & z = 0 \end{cases}$$

satisfy the Cauchy-Riemann equations at $z = 0$, but that they are not differentiable there.

3.6.4 Let f be analytic in a domain D. Show that f is constant on D if and only if each of the following equations holds.

   a. $f'(z) = 0$ for all $z \in D$
   b. Re $f(z) = $ constant   for all $z \in D$
   c. Im $f(z) = $ constant   for all $z \in D$
   d. $|f(z)| = $ constant   for all $z \in D$
   e. Arg $f(z) = $ constant   for all $z \in D$
      or $f(z) = 0$ for all $z \in D$
   f. $\bar{f}$ is analytic in D

3.6.5 Show that if two analytic functions have equal derivatives in a domain D, then they differ by an additive constant in D.

3.6.6 If $f(z) = u(x,y) + iv(x,y)$, show that the curves

$$u(x,y) = \text{constant} \qquad \text{and} \qquad v(x,y) = \text{constant}$$

are orthogonal at every point z where $f'(z)$ exists and is not zero.

3.6.7 Show that the only entire function f that satisfies the initial value problem

$$f'(z) = f(z) \qquad f(0) = 1$$

is $f(z) = e^z$.

3.6.8 Let $f(z) = u(x,y) + iv(x,y)$. Introducing polar coordinates

$$x = r \cos \theta \qquad y = r \sin \theta$$

prove that the Cauchy-Riemann equations at any point $z \neq 0$ in *polar form* are

$$\frac{\partial u}{\partial r} = \frac{1}{r}\frac{\partial v}{\partial \theta} \qquad \frac{\partial v}{\partial r} = -\frac{1}{r}\frac{\partial u}{\partial \theta}$$

Furthermore,

$$f'(z) = (\cos \theta - i \sin \theta)\frac{\partial}{\partial r}[u(r,\theta) + iv(r,\theta)]$$

3.7

HARMONIC FUNCTIONS

Let the function $f(z) = u(x,y) + iv(x,y)$ be analytic in some domain D. Then by Corollary 3.4, the four partial derivatives u_x, u_y, v_x, and v_y exist and satisfy the Cauchy-Riemann equations

$$u_x = v_y \qquad u_y = -v_x$$

at each point z of D. It will be proved in Sec. 4.4 that the existence of f' in D implies the existence and continuity of the partial derivatives of u and v of all orders. In any case, let us assume here (until we prove it in Sec. 4.4) the existence and continuity of the second partial derivatives of u and v. In particular, this implies that

$$u_{xy} = u_{yx} \qquad \text{and} \qquad v_{xy} = v_{yx} \tag{3.10}$$

Differentiating the Cauchy-Riemann equations (3.7) and using (3.10), we obtain

$$u_{xx} + u_{yy} = 0 \qquad \text{and} \qquad v_{xx} + v_{yy} = 0$$

Thus each of the functions u and v must be a solution of the second-order partial differential equation

$$\nabla^2 \phi \equiv \frac{\partial^2 \phi}{\partial x^2} + \frac{\partial^2 \phi}{\partial y^2} = 0 \tag{3.11}$$

which is called the *Laplace* or *potential equation*.

PIERRE SIMON LAPLACE

Pierre Simon Laplace was born in Beaumont-en-Auge, France, on March 23, 1749. His parents were farmers. With the support of D'Alembert, Laplace became a professor at the École Militaire of Paris, where in 1785 he administered Napoleon's entrance examination.

With every political change in France, Laplace's political beliefs also changed. He was always a strong supporter of the party in power and thus he was continually gaining honors, titles, and wealth.

Laplace contributed significantly to the fields of celestial mechanics and probability theory. In 1773 he presented a paper before the Academy of Sciences in which he demonstrated the natural stability of the solar system (Newton had postulated that divine intervention was occasionally necessary!). The equation [see Eq. (3.11)] that bears Laplace's name today had been found by Euler in 1752 in connection with Euler's studies on hydrodynamics.

Laplace died on March 5, 1827, at the age of 77.

This equation is of fundamental importance in many branches of pure and applied mathematics. Among the functions of physical significance which satisfy Laplace's equation are the following: the potential at a point interior to a domain devoid of matter in a gravitational field, the potential at a point interior to a domain devoid of charges in an electric field, and the velocity potential and stream function of a two-dimensional irrotational flow of an incompressible nonviscous fluid.

A function $\phi(x, y)$ whose second partial derivatives exist and are continuous in a domain D is called *harmonic* in D if it satisfies Laplace's equation throughout D. It follows from the previous discussion that if the function $f(z) = u(x, y) + iv(x, y)$ is analytic in D, then u and v are harmonic in D. Therefore, even the real part alone (or the imaginary part) cannot be chosen arbitrarily. Only harmonic functions can serve as the real part (or imaginary part) of an analytic function. Let u be a harmonic function in D. If there exists a harmonic function v in D such that the function $f(z) = u(x, y) + iv(x, y)$ is analytic for every $z = x + iy \in D$, then v is called a *harmonic conjugate* of u. By Theorem 3.6, the harmonic conjugate (if it exists) is unique up to an additive constant. Also, if v is a harmonic conjugate of u, then $-v + iu$ is analytic, and therefore $-u$ is a harmonic conjugate of v.

Example 3.14 Let D be a domain with the following property: There exists a point $z_0 = x_0 + iy_0 \in D$ such that for any point $z = x + iy \in D$, the line segments joining (x_0, y_0) to (x_0, y) and (x_0, y) to (x, y) both lie in D. See Fig. 3.3. For example, disks have this property. Then any function $u(x, y)$ that is harmonic in D has a harmonic conjugate $v(x, y)$ in D.

Proof. Since $f = u + iv$ must be analytic in D, Eq. (3.7) must hold. Integrating the equation $v_x = -u_y$ with respect to x, we get

$$v(x, y) = -\int_{x_0}^{x} u_y(t, y)\, dt + c(y) \tag{3.12}$$

where c is some (continuously differentiable) function of y. Since $v_y = u_x$ and u is harmonic in D, we obtain from Eq. (3.12)

$$u_x(x, y) = -\frac{\partial}{\partial y} \int_{x_0}^{x} u_y(t, y)\, dt + c'(y)$$

$$= -\int_{x_0}^{x} u_{yy}(t, y)\, dt + c'(y)$$

$$= \int_{x_0}^{x} u_{xx}(t, y)\, dt + c'(y)$$

$$= u_x(x, y) - u_x(x_0, y) + c'(y)$$

Hence

$$c'(y) = u_x(x_0, y)$$

Fig. 3.3

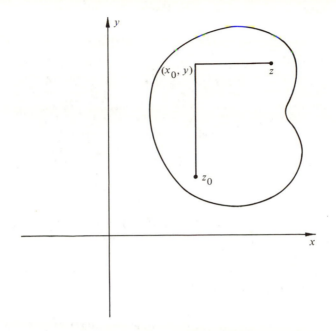

Integrating this equation with respect to y, we obtain

$$c(y) = \int_{y_0}^{y} u_x(x_0, t) \, dt + c$$

where c is some constant. Hence

$$v(x, y) = -\int_{x_0}^{x} u_y(t, y) \, dt + \int_{y_0}^{y} u_x(x_0, t) \, dt + c \qquad (3.13)$$

The work we have done so far shows how to construct v. Now, using formula (3.13) directly, we can easily verify that v is continuously differentiable in x and y and that it satisfies Eq. (3.7). That is, v is a harmonic conjugate of u in D.

Example 3.15 Show that the function $u(x, y) = e^x \cos y$ is harmonic in $\mathbf{C}$, and construct a harmonic conjugate for u.

Solution. Clearly, the second partial derivatives of u exist, are continuous, and satisfy Laplace's equation

$$\nabla^2 u \equiv u_{xx} + u_{yy} = e^x \cos y - e^x \cos y = 0$$

in $\mathbf{C}$. Hence u is harmonic in $\mathbf{C}$. To construct a harmonic conjugate v of u in $\mathbf{C}$ we either use formula (3.13) with (x_0, y_0) being any point, say $(0,0)$, or we apply the steps of Example 3.14. We recommend the latter approach. We must have $v_x = -u_y = e^x \sin y$. Integrating with respect to x, we get

$$v(x,y) = e^x \sin y + c(y)$$

But $v_y = u_x$. That is,

$$e^x \cos y + c'(y) = e^x \cos y$$

and therefore $c'(y) = 0$; that is, $c(y) = c$ (constant). Hence

$$v(x,y) = e^x \sin y + c$$

For $c = 0$ we get the analytic function

$$f(z) = e^x \cos y + ie^x \sin y = e^z$$

Example 3.16 Show that the function $u(x,y) = x^2 - y^2 + x - y$ is harmonic in $\mathbf{C}$ and construct its harmonic conjugates.

Solution. The reader can easily verify that u is harmonic in $\mathbf{C}$. Using the Cauchy-Riemann equations, we obtain

$$u_x = 2x + 1 = v_y \qquad \text{and} \qquad u_y = -2y - 1 = -v_x$$

Integrating these equations with respect to y and x, respectively, we obtain

$$v(x,y) = 2xy + y + c_1(x) \qquad \text{and} \qquad v(x,y) = 2xy + x + c_2(y)$$

Thus $y + c_1(x) = x + c_2(y)$, which implies that

$$c_1(x) - x = c_2(y) - y = c \qquad \text{for some } c \in \mathbf{C}$$

Hence the function

$$v(x,y) = 2xy + x + y + c$$

is a harmonic conjugate of u, and clearly all harmonic conjugates of u arise this way.

EXERCISES

3.7.1 Compute a harmonic conjugate for the following harmonic functions.
   a. $u = x^3 - 3xy^2 + 1$ b. $u = e^x \sin y$
   c. $u = xe^x \cos y - ye^x \sin y$
3.7.2 Choose the constant a so that the function $u(x,y) = x^2 + ay^2$ is harmonic in $\mathbf{C}$. Then compute the harmonic conjugate v of u with the additional property that $v(0,0) = 3$.
3.7.3 Show that the function $u(x,y) = \log \sqrt{x^2 + y^2}$ is harmonic in $\mathbf{C} - \{0\}$ and a harmonic conjugate of $u(x,y)$ in $\mathbf{C} - \{z \in \mathbf{C}: \operatorname{Re} z \le 0\}$ is $\operatorname{Arg} z$.

3.7.4 Show that a harmonic conjugate of the harmonic function

$$u(x,y) = \frac{x}{x^2 + y^2}$$

with $(x,y) \neq (0,0)$, is

$$v(x,y) = -\frac{y}{x^2 + y^2}$$

3.7.5 If u and v are harmonic in a domain D, show that the function $(u_y - v_x) + i(u_x + v_y)$ is analytic in D.

3.7.6 Let f be analytic in a domain D that does not include the point 0, and for $z = re^{i\theta}$ write $f(z) = u(r,\theta) + iv(r,\theta)$, where u and v have continuous second partial derivatives. Show that u and v satisfy Laplace's equation in *polar form*:

$$\frac{\partial^2 \phi}{\partial r^2} + \frac{1}{r}\frac{\partial \phi}{\partial r} + \frac{1}{r^2}\frac{\partial^2 \phi}{\partial \theta^2} = 0$$

3.7.7 Let $f(z) = u(x,y) + iv(x,y)$ be an entire function. Construct f in terms of $u + v$. In particular, show that if $u(x,y) + v(x,y) = e^x \cos y + e^x \sin y$, then there exists $\lambda \in \mathbf{C}$ such that $f(z) = e^z + \lambda$ for $z \in \mathbf{C}$.

Chapter 4 Complex Integration

INTRODUCTION

The theory of complex integration is the heart of an elementary course in complex variables. The concept of the complex line integral was introduced by Cauchy in 1814; he published his basic theorems on functions of one complex variable in 1825. As the reader will see, most of the results in this chapter are in fact either due to Cauchy or else are generalizations of his results. Complex integration is of great interest to both pure and applied mathematicians, physicists, and engineers. One advantage of complex integration is that many interesting properties of functions of a complex variable follow very easily when the techniques of complex integration are used, and they can be extremely difficult to prove without such techniques. As elementary examples, let us mention the following facts: an analytic function in a domain D has derivatives of all orders (see Theorem 4.5); a harmonic function in a domain D has all partial derivatives of all orders (see Corollary 4.2); the values of an analytic function in a domain D are uniquely determined by its values in any subset of D containing a limit point in D (see Corollary 4.5). Complex integration also enables us to give a short proof of the fundamental theorem of algebra (see Theorem 4.8) and to prove that if f is a nonconstant analytic function in a domain D, then the maximum of $|f|$ is not attained in D (see Theorem 4.13). Many interesting

AUGUSTIN LOUIS CAUCHY

Augustin Louis Cauchy was born in Paris on August 21, 1789, about six weeks after the fall of the Bastille. His father was a parliamentary lawyer and a police lieutenant in Paris when the Bastille fell. Fear of the guillotine forced the Cauchy family to move to the village of Arcueil, where they spent eleven years in near starvation. Due to this early malnutrition, Cauchy's health was never good. On January 1, 1800, Augustin's father, who had kept his connections in Paris, was elected Secretary of the Senate.

Cauchy received an excellent education. His father undertook his education until he was 13. Then Augustin entered the Central School of the Panthéon, where he excelled in Greek, Latin, and humanities. At 16, he passed second into the École Polytechnique in Paris, and from there he went to a civil engineering school. Upon completion of his training in 1810, he served under Napoleon as a military engineer in Cherbourg for three years.

By 1816, his mathematical work had brought him so much fame that he was elected to the Academy of Sciences and was made a full professor at the École Polytechnique. He also received appointments to the College de France and the Sorbonne.

Cauchy was married in 1818 to Aloise de Bure, with whom he lived happily for about 40 years. Two daughters were born of their marriage.

Cauchy was a prolific writer, surpassed only by Euler. He published a total of 789 papers on all branches of mathematics. One of his favorite topics was the theory of complex functions, a field in which he published continuously after 1814. Cauchy is considered the founder of modern analysis. However, during his life he was often criticized because his vote for candidates in academic and scientific societies was influenced by his political and religious beliefs (ardently pro-monarchist and pro-Catholic).

During the first eight years of the July monarchy (1830–1848), Cauchy lived in voluntary exile because of his political beliefs. Upon his return to France, he held only positions that did not require an oath of allegiance to the government. Finally, in 1848, the 59-year-old Cauchy was able to return to the professorship offered him 32 years earlier at the École Polytechnique. He died from bronchitis on May 23, 1857.

real integrals are evaluated using the techniques of complex integration (see Examples 4.4, 4.5, 4.9, and 4.10). The Taylor-series expansion of analytic functions is derived in this chapter. Cauchy's theorem and Cauchy's integral formula are established here in a very precise, though elementary, way.

We should remark here that one of the key tools in our approach in this chapter is the fact that every analytic function is locally a power series. See Corollary 4.3. Our approach is essentially due to Weierstrass (and in fact, Weierstrass defined an analytic function to be locally a power series).

4.2

CONTOUR INTEGRATION

Let $a \leq b$, and let $\gamma \colon [a,b] \to \mathbf{C}$ be a continuous function. Recall that we called γ a *curve*. We call $\{\gamma(t) \colon a \leq t \leq b\}$ the *trace* of the curve γ. If $A \subset \mathbf{C}$ and the trace of γ is contained in A, then γ is called a *curve in A*. Note that two curves may have the same trace and yet be different curves. For example, the curves $\gamma_1(t) = e^{2\pi i t}$ and $\gamma_2(t) = e^{-4\pi i t}$, with $0 \leq t \leq 1$, both have the circle $|z| = 1$ as trace, but γ_1 traces out the circle $|z| = 1$ exactly once in the counterclockwise direction, whereas γ_2 traces out the circle $|z| = 1$ twice in the clockwise direction.

The curve γ is called *simple* if $\gamma(t_1) \neq \gamma(t_2)$ whenever $a < t_1 < t_2 < b$, while it is called *closed* if $\gamma(a) = \gamma(b)$.

The inverse of γ is the curve $\tilde{\gamma} \colon [a,b] \to \mathbf{C}$ defined by

$$\tilde{\gamma}(t) = \gamma(a + b - t) \qquad a \leq t \leq b$$

Clearly $\tilde{\gamma}$ has the same trace as γ; however, $\tilde{\gamma}$ is said to have the opposite orientation of γ. Let $\gamma_1 \colon [a_1,b_1] \to \mathbf{C}$ and $\gamma_2 \colon [a_2,b_2] \to \mathbf{C}$ be curves with $\gamma_1(b_1) = \gamma_2(a_2)$. Then the *sum* $(\gamma_1 + \gamma_2)$ of these two curves is the curve $\gamma_1 + \gamma_2 \colon [a_1, b_1 + b_2 - a_2] \to \mathbf{C}$ defined by

$$(\gamma_1 + \gamma_2)\,(t) = \begin{cases} \gamma_1(t) & a_1 \leq t \leq b_1 \\ \gamma_2(t + a_2 - b_1) & b_1 \leq t \leq b_1 + b_2 - a_2 \end{cases}$$

If $\gamma_1, \ldots, \gamma_n$ are curves with $\gamma_i(b_i) = \gamma_{i+1}(a_{i+1})$ for $i = 1, 2, \ldots, n - 1$, we define the curve $\gamma_1 + \cdots + \gamma_n$ inductively by the formula $\gamma_1 + \cdots + \gamma_n = (\gamma_1 + \cdots + \gamma_{n-1}) + \gamma_n$.

If $\gamma(t) = x(t) + iy(t)$ for $a \leq t \leq b$, then we define the *derivative* of γ by $\gamma'(t) = x'(t) + iy'(t)$, where the derivatives at a and b are understood to be one-sided derivatives.

If there exists a partition $a = t_0 < t_1 < \cdots < t_n = b$ of $[a,b]$ such that the curve $\gamma \colon [a,b] \to \mathbf{C}$ when restricted to any closed subinterval $[t_{j-1}, t_j]$, $j = 1, \ldots, n$, has a continuous derivative, then we call γ a *contour* (or *piecewise differentiable curve*).

As in the case of the derivative γ' of γ, we define the *complex integral* of γ to be

$$\int_a^b \gamma(t)\,dt = \int_a^b x(t)\,dt + i \int_a^b y(t)\,dt$$

KARL THEODOR WEIERSTRASS

Karl Theodor Weierstrass was born at Ostenfelde, Germany, on October 31, 1815. His father was a customs officer in the pay of the French (who at the time dominated Europe). His mother died when he was 11 years old, and his stepmother, to say the least, contributed very little to Karl's education.

Karl not only did extremely well at school but at the age of 15 was able to secure a job as an accountant. After analyzing Karl's qualities, his father concluded that his son should be prepared for public service and in order that his son should effectively "pluck" and never "be plucked," he determined that Karl should study law at the University of Bonn. After four years at Bonn, Karl returned home expert in drinking and fencing, but without the law degree.

Following the advice of a family friend, Karl prepared for secondary school teaching at the Academy of Münster, from which he graduated at the age of 26. At Münster, Weierstrass became fascinated by the lectures of his mathematics professor, Christop Gudermann, who was an enthusiast of elliptic functions. Gudermann's idea was to base everything on the power series representation of a function. This idea was the main tool for the greatest part of Weierstrass's work.

For fifteen years Weierstrass taught at various secondary schools. His paper on abelian functions (published in 1854) brought Weierstrass so much recognition that ten years later he was a professor at the University of Berlin. His perfectly organized lectures enjoyed great popularity. It is mainly through these lectures that Weierstrass's ideas became the common property of all mathematicians. In 1856 he suffered a nervous breakdown, but he continued to be mathematically active to the end of his life. Weierstrass contributed to all fields of analysis, and he was undoubtedly one of the founders of complex function theory.

Although Weierstrass never married, he liked women. The famous (and beautiful) Russian mathematician Sonja Kowalewski was his student and his "weakness." He died from influenza on February 19, 1897.

Let $\gamma: [a,b] \to \mathbf{C}$ be a contour, and let f be a complex-valued function such that $f \circ \gamma$ is continuous. We then call f a *continuous complex-valued function on the contour* γ, and we define the *contour integral of f along the contour* γ by the formula

$$\int_\gamma f(z) \, dz = \int_a^b f(\gamma(t))\gamma'(t) \, dt$$

An equivalent definition of the contour integral $\int_\gamma f(z) \, dz$ is the following: Let $a = t_0 < t_1 < \cdots < t_n = b$ be a partition of $[a,b]$. Choose a point $\tau_k \in [t_{k-1}, t_k]$ for $k = 1, 2, \ldots, n$ and set $z_k = \gamma(t_k)$ and $\zeta_k = \gamma(\tau_k)$ for $k = 1, 2, \ldots, n$. Form the Riemann sum

$$I_n = \sum_{k=1}^n f(\zeta_k)(z_k - z_{k-1})$$

We leave it as an exercise for the reader to show that I_n approaches $\int_\gamma f(z) \, dz$ as $n \to \infty$ and as $\max_k (t_k - t_{k-1})$ approaches zero.

The *length* of the contour γ, $L(\gamma)$, is defined by

$$L(\gamma) = \int_a^b |\gamma'(t)| \, dt$$

Clearly the above integral exists, as $|\gamma'|$ is piecewise continuous.

If $c \leq d$ and $s: [c,d] \to [a,b]$ is a piecewise differentiable curve such that $s(c) = a$ and $s(d) = b$, we call $\gamma \circ s$ a *reparametrization* of γ. [In the literature, the requirement is often made that $s'(t)$ be positive for all $t \in [c,d]$. The reader should be aware that we do not impose such a restriction in our definition.]

In the following theorem we develop some useful properties of contour integration.

Theorem 4.1 Let f and g be continuous complex-valued functions defined on the contour $\gamma: [a,b] \to \mathbf{C}$. Then

a. For any pair of complex numbers k and l,

$$\int_\gamma [kf(z) + lg(z)] \, dz = k \int_\gamma f(z) \, dz + l \int_\gamma g(z) \, dz$$

That is, complex integration is a linear operation.

b. Changing the orientation of γ reverses the sign of the integral; that is,

$$\int_\gamma f(z) \, dz = - \int_{\tilde{\gamma}} f(z) \, dz$$

c. If γ_1 and γ_2 are contours and $\gamma = \gamma_1 + \gamma_2$, then

$$\int_\gamma f(z) \, dz = \int_{\gamma_1} f(z) \, dz + \int_{\gamma_2} f(z) \, dz$$

d. $\left| \int_{\gamma} f(z) \, dz \right| \le \int_a^b |f(\gamma(t))\gamma'(t)| \, dt.$

e. If $\gamma \circ s$ is a reparameterization of γ, then

$$\int_{\gamma \circ s} f(z) \, dz = \int_{\gamma} f(z) \, dz$$

f. If $|f(\gamma(t))| \le M$ for $a \le t \le b$ and L is the length of the contour γ, then

$$\left| \int_{\gamma} f(z) \, dz \right| \le ML$$

Proof
a. By the definition of the integral, we have

$$\int_{\gamma} [kf(z) + lg(z)] \, dz = \int_a^b [kf(\gamma(t)) + lg(\gamma(t))]\gamma'(t) \, dt$$

$$= \int_a^b [kf(\gamma(t))\gamma'(t) + lg(\gamma(t))\gamma'(t)] \, dt$$

$$= k \int_a^b f(\gamma(t))\gamma'(t) \, dt + l \int_a^b g(\gamma(t))\gamma'(t) \, dt$$

$$= k \int_{\gamma} f(z) \, dz + l \int_{\gamma} g(z) \, dz$$

b. Recall that $\tilde{\gamma}(t) = \gamma(a + b - t)$, and so $\tilde{\gamma}'(t) = -\gamma'(a + b - t)$. Hence

$$\int_{\tilde{\gamma}} f(z) \, dz = \int_a^b f(\tilde{\gamma}(t))\tilde{\gamma}'(t) \, dt$$

$$= -\int_a^b f(\gamma(a + b - t))\gamma'(a + b - t) \, dt$$

$$= -\int_a^b f(\gamma(t))\gamma'(t) \, dt$$

$$= -\int_{\gamma} f(z) \, dz$$

c. Let $\gamma_1 : [a, b_1] \to \mathbf{C}$ and $\gamma_2 : [a_2, b_2] \to \mathbf{C}$ be curves such that $\gamma = \gamma_1 + \gamma_2$.

Then $b = b_1 + b_2 - a_2$, and

$$\int_\gamma f(z)\, dz = \int_a^b f(\gamma(t))\gamma'(t)\, dt$$

$$= \int_a^{b_1} f(\gamma(t))\gamma'(t)\, dt + \int_{b_1}^b f(\gamma(t))\gamma'(t)\, dt$$

$$= \int_a^{b_1} f(\gamma_1(t))\gamma_1'(t)\, dt$$

$$+ \int_{b_1}^b f(\gamma_2(t + a_2 - b_1))\gamma_2'(t + a_2 - b_1)\, dt$$

$$= \int_a^{b_1} f(\gamma_1(t))\gamma_1'(t)\, dt + \int_{a_2}^{b_2} f(\gamma_2(t))\gamma_2'(t)\, dt$$

$$= \int_{\gamma_1} f(z)\, dz + \int_{\gamma_2} f(z)\, dz$$

d. If $\int_\gamma f(z)\, dz = 0$, we are finished. So assume that $\int_\gamma f(z)\, dz \neq 0$. Then if $\theta \in \mathbf{R}$,

$$\mathrm{Re}\left[e^{-i\theta} \int_\gamma f(z)\, dz \right] = \int_a^b \mathrm{Re}\left[e^{-i\theta} f(\gamma(t))\gamma'(t) \right] dt$$

$$\leq \int_a^b |f(\gamma(t))\gamma'(t)|\, dt$$

So by choosing $\theta \in \arg\left[\int_\gamma f(z)\, dz \right]$ (and recalling that $z = |z|e^{i\,\arg z}$), the result follows.

e. By the fundamental theorem of calculus, it follows that there exists a continuous function $F\colon [a,b] \to \mathbf{C}$ such that except for finitely many $t \in [a,b]$, $F'(t) = f(\gamma(t))\gamma'(t)$. So

$$\int_\gamma f(z)\, dz = \int_a^b f(\gamma(t))\gamma'(t)\, dt = F(b) - F(a)$$

$$= F(s(d)) - F(s(c))$$

$$= \int_c^d (F \circ s)'(t)\, dt$$

$$= \int_c^d F'(s(t))s'(t)\, dt$$

$$= \int_c^d f(\gamma(s(t)))\gamma'(s(t))s'(t)\, dt$$

$$= \int_c^d f(\gamma \circ s(t))(\gamma \circ s)'(t)\, dt$$

$$= \int_{\gamma \circ s} f(z)\, dz$$

f. $\left| \int_\gamma f(z) \, dz \right| \leq \int_a^b |f(\gamma(t))\gamma'(t)| \, dt \leq M \int_a^b |\gamma'(t)| \, dt = ML$

Remark 4.1 We shall often speak of a simple closed contour tracing out a circle, ellipse, triangle, or rectangle as if it were its trace. It will always be understood that the contour in question is simple and is parametrized in the counterclockwise direction. The justification for this convention follows from Theorem 4.1c and e; that is, the integral of a complex-valued function over a simple closed contour is independent of the particular choice of parametrization and is also independent of the choice of the common initial and final point.

Example 4.1 If $f: D \to \mathbf{C}$ has a continuous derivative and γ is a contour in D, then the fundamental theorem of calculus holds. That is,

$$\int_\gamma f'(z) \, dz = f|_\gamma$$

where $f|_\gamma$ is called the *variation* of f along γ and is defined to be $f|_\gamma = f(\gamma(b)) - f(\gamma(a))$. Thus in this case, the integral is independent of the contour of integration and depends only on the endpoints of the contour.

Proof

$$\int_\gamma f'(z) \, dz = \int_a^b f'(\gamma(t))\gamma'(t) \, dt$$

$$= \int_a^b (f \circ \gamma)'(t) \, dt$$

$$= f(\gamma(b)) - f(\gamma(a))$$

Example 4.2 Let $\gamma(t) = z_0 + re^{it}$, with $a \leq t \leq b$. Show that

$$\frac{1}{2\pi i} \int_\gamma \frac{dz}{z - z_0} = \frac{b - a}{2\pi}$$

Proof

$$\frac{1}{2\pi i} \int_\gamma \frac{dz}{z - z_0} = \frac{1}{2\pi i} \int_a^b \frac{rie^{it}}{re^{it}} \, dt = \frac{b - a}{2\pi}$$

If in the above integral the curve γ is closed, that is, if $e^{ia} = e^{ib}$ (or, in other words, $b - a = 2\pi k$ for some integer k), then

$$\frac{1}{2\pi i} \int_\gamma \frac{dz}{z - z_0} = k \qquad \text{for some } k \in \mathbf{Z}$$

In this case, k is called the *winding number* of γ with respect to the point z_0. Geometrically it represents the number of times the point z goes around the point z_0 as z traverses the closed curve γ.

Example 4.3 Evaluate the integral $\int_\gamma (2\bar{z} - 1)\, dz$ where

a. $\gamma = \overline{1, -i}$ is the straight-line segment joining 1 to $-i$
b. γ is the arc in the second quadrant along the circle $|z| = 1$ joining i to -1 given by $\gamma(t) = e^{2\pi i t}$ for $1/4 \le t \le 1/2$
c. γ is the closed triangle $\overline{0, 1, i, 0}$

Solution
a. Set $\gamma(t) = 1 + (-i - 1)t$, for $0 \le t \le 1$. Hence

$$\int_\gamma (2\bar{z} - 1)\, dz = \int_0^1 (1 + 2it - 2t)(-i - 1)\, dt$$

$$= \int_0^1 (4t - 1 - i)\, dt = 2t^2 - (1 + i)t \big|_0^1$$

$$= 2 - (1 + i) = 1 - i$$

b. We have $\gamma(t) = e^{2\pi i t}$ if $1/4 \le t \le 1/2$. So

$$\int_\gamma (2\bar{z} - 1)\, dz = \int_{1/4}^{1/2} (2e^{-2\pi i t} - 1)2\pi i e^{2\pi i t}\, dt$$

$$= 1 + (2\pi + 1)i$$

c. The three sides are represented by $\gamma_1 = \overline{0,1}$, $\gamma_2 = \overline{1,i}$, and $\gamma_3 = \overline{i,0}$. Thus

$$\int_\gamma (2\bar{z} - 1)\, dz = \int_{\gamma_1} (2\bar{z} - 1)\, dz + \int_{\gamma_2} (2\bar{z} - 1)\, dz$$

$$+ \int_{\gamma_3} (2\bar{z} - 1)\, dz$$

$$= 0 + (1 + i) + (-1 + i) = 2i$$

EXERCISES
4.2.1 Consider the contours

$$\gamma_1(t) = z_0 e^{it} \qquad 0 \le t \le \pi/4$$
$$\gamma_2(t) = z_0 e^{-it} \qquad 0 \le t \le 7\pi/4$$

Compute the sum $\gamma_1 + \tilde{\gamma}_2$.

4.2.2 Consider the contours with initial point 0 and terminal point $1 + i$ given by

$$\gamma_1(t) = t + it \qquad \text{for } 0 \le t \le 1 \text{ (straight-line segment)}$$

$$\gamma_2(t) = (1 - \cos t) + i \sin t \qquad \text{for } 0 \le t \le \pi/2$$
$$\text{(a quadrant of a circle)}$$

$$\gamma_3(t) = t \qquad \text{for } 0 \le t \le 1$$

and

$$\gamma_3(t) = 1 + i(t - 1) \qquad \text{for } 1 \le t \le 2 \text{ (a broken-line path)}$$

Prove the following.

a. $\displaystyle\int_{\gamma_1} x \, dz = \frac{1 + i}{2}$
b. $\displaystyle\int_{\gamma_2} x \, dz = \tfrac{1}{2} + i(1 - \pi/4)$

c. $\displaystyle\int_{\gamma_3} x \, dz = \tfrac{1}{2} + i$
d. $\displaystyle\int_{\gamma_1} y \, dz = \frac{1 + i}{2}$

e. $\displaystyle\int_{\gamma_2} y \, dz = \frac{\pi}{4} + \frac{i}{2}$
f. $\displaystyle\int_{\gamma_3} y \, dz = \frac{i}{2}$

g. $\displaystyle\int_{\gamma_1} z \, dz = \int_{\gamma_2} z \, dz = \int_{\gamma_3} z \, dz = i$

h. $\displaystyle\int_{\gamma_1} \bar{z} \, dz = 1$
i. $\displaystyle\int_{\gamma_2} \bar{z} \, dz = 1 + i(1 - \pi/2)$

j. $\displaystyle\int_{\gamma_3} \bar{z} \, dz = 1 + i$

4.2.3 Let γ be a contour with initial point z_1 and final point z_2. Show that the following are true.

a. $\displaystyle\int_{\gamma} e^z \, dz = e^{z_2} - e^{z_1}$

b. $\displaystyle\int_{\gamma} \cos z \, dz = \sin z_2 - \sin z_1$

c. $\displaystyle\int_{\gamma} z^n \, dz = \frac{z_2^{n+1} - z_1^{n+1}}{n + 1}, \quad n \ne -1$

assuming that in the case of $n < 0$, $\gamma(t) \ne 0$ for all $t \in [a,b]$.

4.2.4 Let f and g be analytic in a domain D with continuous derivatives in D. Let γ be a contour in D. Then

$$\int_{\gamma} f(z)g'(z) \, dz = fg|_{\gamma} - \int_{\gamma} f'(z)g(z) \, dz$$

This is the integration-by-parts formula.

4.2.5 Given that γ is the circle $|z| = 1$, show that

$$\left| \int_\gamma \frac{dz}{4 + 3z} \right| \leq \frac{6}{5} \pi$$

4.2.6 Let $\gamma(t) = e^{-4\pi it}$ for $0 \leq t \leq 1$. Evaluate the integrals

$$\int_\gamma \frac{dz}{z} \qquad \int_\gamma \frac{dz}{z^3} \qquad \int_\gamma \frac{dz}{|z|}$$

4.3
CAUCHY'S THEOREM FOR STAR DOMAINS

We shall first prove Cauchy's theorem for a triangle, and then for a star domain. See Cauchy's biography.

Theorem 4.2 Let f be analytic in the interior and on the perimeter of a triangle T. Then

$$\int_\gamma f(z)\, dz = 0$$

where γ is the contour describing the triangle T in the counterclockwise direction.

Proof. By joining the midpoints of the sides of triangle T by line segments, we obtain four triangles T_a, T_b, T_c, T_d with respective boundaries γ_a, γ_b, γ_c, γ_d traversed in the positive direction. See Fig. 4.1. Then since the integrals taken

Fig. 4.1

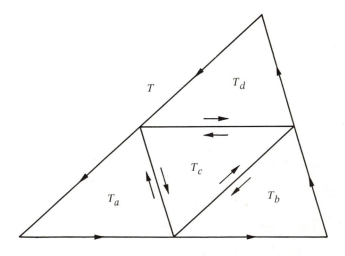

in opposite directions along the sides of T_c are of opposite sign (see Theorem 4.1b), we have

$$\int_\gamma f(z)\,dz = \int_{\gamma_a} f(z)\,dz + \int_{\gamma_b} f(z)\,dz + \int_{\gamma_c} f(z)\,dz + \int_{\gamma_d} f(z)\,dz \quad (4.1)$$

From (4.1) it follows that

$$\left| \int_\gamma f(z)\,dz \right| \le \left| \int_{\gamma_a} f(z)\,dz \right| + \left| \int_{\gamma_b} f(z)\,dz \right| + \left| \int_{\gamma_c} f(z)\,dz \right|$$

$$+ \left| \int_{\gamma_d} f(z)\,dz \right| \quad (4.2)$$

Of the four nonnegative numbers appearing on the right side of (4.2), one must be larger than or equal to the other three. Denote by T_1 a triangle corresponding to that number and by γ_1 its boundary. Then

$$\left| \int_\gamma f(z)\,dz \right| \le 4 \left| \int_{\gamma_1} f(z)\,dz \right|$$

By joining the midpoints of the sides of triangle T_1 by line segments, we obtain four new triangles. Proceeding with T_1 as with T, we find a triangle T_2 with boundary γ_2 such that

$$\left| \int_{\gamma_1} f(z)\,dz \right| \le 4 \left| \int_{\gamma_2} f(z)\,dz \right|$$

Thus

$$\left| \int_\gamma f(z)\,dz \right| \le 4^2 \left| \int_{\gamma_2} f(z)\,dz \right|$$

Continuing in this manner, we get a sequence of nested triangles

$$T \supset T_1 \supset T_2 \supset \cdots T_n \supset \cdots \quad (4.3)$$

with corresponding boundaries $\gamma, \gamma_1, \gamma_2, \ldots, \gamma_n, \ldots$ such that

$$\left| \int_\gamma f(z)\,dz \right| \le 4^n \left| \int_{\gamma_n} f(z)\,dz \right| \quad (4.4)$$

Clearly if L is the length of γ and L_n is the length of γ_n, then

$$L_n = \frac{1}{2^n} L \quad (4.5)$$

So as $d(T_n) \le L_n$, we see that $d(T_n) \to 0$ as $n \to \infty$.

Hence, in view of Eq. (4.5), the nested sequence of closed sets in (4.3) satisfies the conditions of Cantor's intersection theorem 2.6. Consequently

$$\bigcap_{n=1}^{\infty} T_n = \{z_0\} \qquad \text{for some } z_0 \in T$$

Since f is analytic at z_0, clearly

$$f(z) = f(z_0) + (z - z_0)f'(z_0) + (z - z_0)\tau(z)$$

where τ is a continuous function satisfying

$$\lim_{z \to z_0} \tau(z) = 0$$

The integral in the right side of Eq. (4.4) then becomes

$$\int_{\gamma_n} f(z)\, dz = f(z_0) \int_{\gamma_n} dz + f'(z_0) \int_{\gamma_n} (z - z_0)\, dz + \int_{\gamma_n} (z - z_0)\tau(z)\, dz$$

Since γ_n is closed, we have from Example 4.1

$$\int_{\gamma_n} dz = 0 \quad \text{and} \quad \int_{\gamma_n} (z - z_0)\, dz = \left. \frac{(z - z_0)^2}{2} \right|_{\gamma_n} = 0$$

Hence
$$\int_{\gamma_n} f(z)\, dz = \int_{\gamma_n} (z - z_0)\tau(z)\, dz \tag{4.6}$$

Since $z_0 \in T_n$, clearly $|z - z_0| \le L_n$ for $z \in \gamma_n$. Let $\varepsilon > 0$. Then there exists $\delta > 0$ such that $|\tau(z)| < \varepsilon$ as long as $|z - z_0| < \delta$. Hence there exists a positive integer N such that if $n \ge N$, then T_n lies inside the disk $|z - z_0| < \delta$. Thus from Eqs. (4.6) and (4.5) we obtain (with $n \ge N$)

$$\left| \int_{\gamma_n} f(z)\, dz \right| \le \varepsilon L_n^2 = \frac{\varepsilon L^2}{4^n}$$

Using this estimate in Eq. (4.4), we get

$$\left| \int_{\gamma} f(z)\, dz \right| \le \varepsilon L^2 \tag{4.7}$$

Since ε is arbitrary, the integral in (4.7) must be zero. The proof is complete.

Next we shall prove Cauchy's theorem for a star domain D. We recall that a domain D is called star with respect to a point $z_0 \in D$ if $z \in D$ implies that $\overline{z_0 z}$ lies in D. For example, the interiors of a circle and a rectangle are star domains.

Theorem 4.3 If f is analytic in a star domain D and γ is a closed contour lying in D, then

$$\int_{\gamma} f(z)\, dz = 0$$

The proof of this theorem is an elementary application of the following lemma and Example 4.1.

Lemma 4.1 If f is analytic in a star domain D, then there exists a function F analytic in D such that

$$F'(z) = f(z)$$

Proof. Let D be star with respect to z_0. Then if z is a point in D, the line segment $\overline{z_0 z}$ lies in D. Consider the function

$$F(z) = \int_{z_0}^{z} f(\zeta)\, d\zeta \qquad \text{for } z \in D \tag{4.8}$$

where by $\int_{z_0}^{z} f(\zeta)\, d\zeta$, we mean $\int_{\overline{z_0 z}} f(\zeta)\, d\zeta$. Let $z \in D$. We shall show that F is differentiable at z. Since $z \in D$, there exists a neighborhood of z contained in D. Thus if h is a complex number with $|h|$ sufficiently small, then $z + h$ is in D and the line segment from z to $z + h$ also lies in D. Clearly the line segment from $z + h$ to z_0 is in D, since D is star with respect to z_0. As D is star with respect to z_0, clearly the triangle $\overline{z_0, z_0 + h, z, z_0}$ together with its interior lies in D. Thus from Theorem 4.2 we have

$$\int_{z_0}^{z+h} f(\zeta)\, d\zeta + \int_{z+h}^{z} f(\zeta)\, d\zeta + \int_{z}^{z_0} f(\zeta)\, d\zeta = 0 \tag{4.9}$$

From Eqs. (4.8) and (4.9) we obtain

$$F(z + h) - F(z) = \int_{z_0}^{z+h} f(\zeta)\, d\zeta - \int_{z_0}^{z} f(\zeta)\, d\zeta = \int_{z}^{z+h} f(\zeta)\, d\zeta$$

Thus for $h \neq 0$, we have

$$\frac{F(z + h) - F(z)}{h} - f(z) = \frac{1}{h} \int_{z}^{z+h} [f(\zeta) - f(z)]\, d\zeta$$

Since f is continuous [the analyticity of f is needed only to invoke Theorem 4.2 in Eq. (4.9), and elsewhere only its continuity is used], it follows that given $\varepsilon > 0$, there exists a $\delta > 0$ such that $|\zeta - z| < \delta$ implies

$$|f(\zeta) - f(z)| < \varepsilon$$

Thus for $|h| < \delta$, we have

$$\left| \frac{F(z + h) - F(z)}{h} - f(z) \right| \leq \frac{1}{|h|}\, \varepsilon |h| = \varepsilon$$

Since ε is arbitrary, the result follows.

Example 4.4 By using the fact that $\int_0^\infty e^{-r^2}\, dr = \sqrt{\pi}/2$, show the following.

a. $\displaystyle\int_0^\infty e^{-x^2} \cos 2ax\, dx = e^{-a^2} \frac{\sqrt{\pi}}{2}$

b. $\displaystyle\int_0^\infty e^{-x^2} \sin 2ax\, dx = e^{-a^2} \int_0^a e^{x^2}\, dx$

Proof. Without loss of generality we assume that $a > 0$. Let

$$\gamma_n = \overline{0, n, n + ia, ia, 0}$$

be the contour of Fig. 4.2. Applying Theorem 4.3 to the entire function e^{-z^2}, we obtain

$$\int_{\gamma_n} e^{-z^2}\, dz = 0$$

Parametrizing the four sides of this rectangle, we see that

$$0 = \int_0^n e^{-x^2}\, dx + \int_0^a ie^{-(n+iy)^2}\, dy$$

$$+ \int_n^0 e^{-(x+ia)^2}\, dx + \int_a^0 ie^{y^2}\, dy$$

$$\equiv J_1 + J_2 + J_3 + J_4 \tag{4.10}$$

Note that

$$J_1 \to \int_0^\infty e^{-x^2}\, dx \qquad \text{as } n \to \infty$$

$$|J_2| \le \int_0^a e^{-n^2+y^2}\, dy = e^{-n^2} \int_0^a e^{y^2}\, dy \to 0 \qquad \text{as } n \to \infty$$

$$J_3 = \int_n^0 e^{-x^2+a^2} (\cos 2ax - i \sin 2ax)\, dx$$

$$= -e^{a^2} \int_0^n e^{-x^2} \cos 2ax\, dx + ie^{a^2} \int_0^n e^{-x^2} \sin 2ax\, dx$$

$$\to -e^{a^2} \int_0^\infty e^{-x^2} \cos 2ax\, dx + ie^{a^2} \int_0^\infty e^{-x^2} \sin 2ax\, dx \qquad \text{as } n \to \infty$$

$$J_4 = -i \int_0^a e^{x^2}\, dx$$

Fig. 4.2

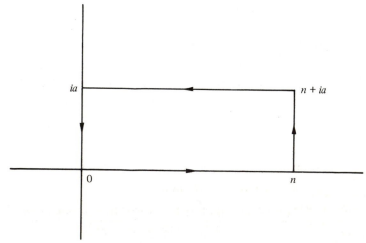

Letting $n \to \infty$ in Eq. (4.10), we obtain

$$\left(\int_0^\infty e^{-x^2} \, dx - e^{a^2} \int_0^\infty e^{-x^2} \cos 2ax \, dx \right)$$

$$- i \left(\int_0^a e^{x^2} \, dx - e^{a^2} \int_0^\infty e^{-x^2} \sin 2ax \, dx \right) = 0$$

The result follows by equating real and imaginary parts.

Example 4.5 For $0 < r < R$ show that the following hold.

a. $\displaystyle\int_0^{2\pi} \frac{d\theta}{R^2 - 2rR \cos \theta + r^2} = \frac{2\pi}{R^2 - r^2}$

b. $\displaystyle\int_0^{2\pi} \frac{\sin \theta \, d\theta}{R^2 - 2rR \cos \theta + r^2} = 0$

c. $\displaystyle\int_0^{2\pi} \frac{\cos \theta \, d\theta}{R^2 - 2rR \cos \theta + r^2} = \frac{r}{R} \frac{2\pi}{R^2 - r^2}$

Proof. Observe that

$$\frac{R + z}{(R - z)z} = \frac{1}{z} + \frac{2}{R - z} \tag{4.11}$$

Integrating (4.11) over the circle $|z| = r$, we obtain

$$\int_0^{2\pi} \frac{R + re^{i\theta}}{(R - re^{i\theta})re^{i\theta}} ire^{i\theta} \, d\theta = \int_{|z|=r} \frac{dz}{z} + \int_{|z|=r} \frac{2}{R - z} \, dz = 2\pi i + 0$$

But

$$\frac{R + re^{i\theta}}{R - re^{i\theta}} = \frac{R^2 - r^2 + 2irR \sin \theta}{R^2 - 2rR \cos \theta + r^2}$$

Hence

$$\int_0^{2\pi} \frac{R^2 - r^2 + 2irR \sin \theta}{R^2 - 2rR \cos \theta + r^2} = 2\pi$$

By equating the real and imaginary parts of this identity, parts (a) and (b) follow. To prove (c), again recall that

$$\int_{|z|=r} \frac{dz}{R - z} = 0$$

and so

$$0 = \int_0^{2\pi} \frac{ire^{i\theta} \, d\theta}{(R - r \cos \theta) - ir \sin \theta}$$

$$= ir \int_0^{2\pi} \frac{(\cos \theta + i \sin \theta)[(R - r \cos \theta) + ir \sin \theta] \, d\theta}{R^2 - 2rR \cos \theta + r^2}$$

Thus

$$\int_0^{2\pi} \frac{(R \cos \theta - r)\, d\theta}{R^2 - 2rR \cos \theta + r^2} + i \int_0^{2\pi} \frac{R \sin \theta\, d\theta}{R^2 - 2rR \cos \theta + r^2} = 0$$

From this identity we get [using part (a) of this exercise]

$$R \int_0^{2\pi} \frac{\cos \theta\, d\theta}{R^2 - 2rR \cos \theta + r^2} = r \int_0^{2\pi} \frac{d\theta}{R^2 - 2rR \cos \theta + r^2} = r \cdot \frac{2\pi}{R^2 - r^2}$$

and (c) follows.

Remark 4.2 Cauchy's theorem and its generalizations, which the reader will encounter in the subsequent sections, were proved by Cauchy in 1814 with the additional assumption that f' is continuous. In 1900 Édouard Goursat (1858–1936, French) proved that it is sufficient to assume that f' exists. The proof of Theorem 4.2, which is due to Alfred Pringsheim (1850–1941, German), is simpler than Goursat's original proof.

EXERCISES

4.3.1 Let f be analytic in the interior and on the perimeter of a rectangle R. Then show that $\int_\gamma f(z)\, dz = 0$, where γ is the contour describing the perimeter of the rectangle R in the counterclockwise direction.

ÉDOUARD JEAN BAPTISTE GOURSAT

Édouard Jean Baptiste Goursat was born at Lanzac, France, on May 21, 1858. He entered the École Normale in 1876 and received his D.Sc. in 1881. One of the best analysts of his time, Goursat was appointed professor of analysis at the University of Paris in 1897. He is best known for his two volumes entitled *Cours d'Analyse Mathématique* (Paris, 1902–1905).

Goursat died in Paris on November 25, 1936.

ALFRED PRINGSHEIM

Alfred Pringsheim was born at Ohlaou, Germany, on September 2, 1850. He is known for his work on the theory of functions. He became a professor at the University of Munich and was author of an encyclopedia of mathematics. Pringsheim died in Zurich, Switzerland, on June 25, 1941.

4.3.2 Show that

$$\int_0^\infty \sin x^2 \, dx = \int_0^\infty \cos x^2 \, dx = \frac{1}{2} \sqrt{\frac{\pi}{2}}$$

This is called *Fresnel's integral* (Augustin Fresnel, 1788–1827, French).

4.4

CAUCHY'S INTEGRAL FORMULA AND APPLICATIONS

A remarkable integral formula due to Cauchy shows that the values of an analytic function on the boundary of a disk determine also the values of the function at all interior points of that disk. From this result it will be shown that an analytic function possesses derivatives of all orders. This clearly is not the case with real-valued functions.

Theorem 4.4 Let f be analytic in the disk $|z - a| < R$, let $0 < r < R$ and $\gamma(t) = a + re^{2\pi i t}$ for $0 \le t \le 1$, and let $|z_0 - a| < r$. Then

$$f(z_0) = \frac{1}{2\pi i} \int_\gamma \frac{f(\zeta)}{\zeta - z_0} \, d\zeta \qquad (4.12)$$

Proof. Clearly there exists a disk centered at z_0 with radius ρ and boundary Γ completely contained in the interior of the disk $|z - a| < r$. See Fig. 4.3.

Let γ_1 be the closed contour $A_1 A_2 A_3 A_4 A_5 A_6 A_1$ and γ_2 be the closed contour $A_1 A_6 A_7 A_4 A_3 A_8 A_1$. Both contours are inside star domains on which the function $f(\zeta)/(\zeta - z_0)$ is analytic. Hence

$$\int_{\gamma_1} \frac{f(\zeta)}{\zeta - z_0} \, d\zeta = 0 \qquad \text{and} \qquad \int_{\gamma_2} \frac{f(\zeta)}{\zeta - z_0} \, d\zeta = 0 \qquad (4.13)$$

Fig. 4.3

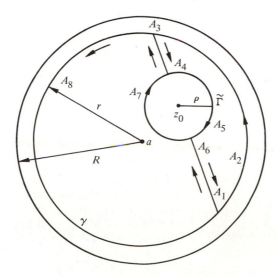

By adding the two integrals in (4.13) and using the fact that the integrals along the straight-line segments common to γ_1 and γ_2 cancel, we get

$$\int_\gamma \frac{f(\zeta)}{\zeta - z_0} \, d\zeta = \int_\Gamma \frac{f(\zeta)}{\zeta - z_0} \, d\zeta \tag{4.14}$$

where Γ is traversed counterclockwise. From (4.14) we obtain

$$\int_\gamma \frac{f(\zeta)}{\zeta - z_0} \, d\zeta = f(z_0) \int_\Gamma \frac{d\zeta}{\zeta - z_0} + \int_\Gamma \frac{f(\zeta) - f(z_0)}{\zeta - z_0} \, d\zeta$$

$$= 2\pi i f(z_0) + \int_\Gamma \frac{f(\zeta) - f(z_0)}{\zeta - z_0} \, d\zeta \tag{4.15}$$

Since f is continuous at z_0, given $\varepsilon > 0$, there exists a $\delta > 0$ such that $|\zeta - z_0| < \delta$ implies

$$|f(\zeta) - f(z_0)| < \varepsilon$$

If in addition the radius of Γ is chosen to be equal to a number $\rho < \delta$, then from Eq. (4.15) we get

$$\left| \int_\gamma \frac{f(\zeta)}{\zeta - z_0} \, d\zeta - 2\pi i f(z_0) \right| \le \frac{\varepsilon}{\rho} 2\pi\rho = 2\pi\varepsilon$$

Since ε is arbitrary, the desired result follows.

Theorem 4.5 Let f be analytic in a (not necessarily star) domain D. Then all the derivatives of f exist and are analytic functions in D.

Proof. Let z_0 be a point in D. Then it will be sufficient to show that $f''(z_0)$ exists. Let $r > 0$ such that the closed disk $|z - z_0| \le r$ is contained in D. Let $\gamma(t) = z_0 + re^{it}$ for $0 \le t \le 2\pi$. See Fig. 4.4. Then if $h \ne 0$ is sufficiently small so that $z_0 + h$ is also in the interior of the disk bounded by γ, we see that by applying Eq. (4.12) to f at z_0 and $z_0 + h$, we can obtain

$$\frac{f(z_0 + h) - f(z_0)}{h} = \frac{1}{2\pi i h} \int_\gamma f(\zeta) \left(\frac{1}{\zeta - z_0 - h} - \frac{1}{\zeta - z_0} \right) d\zeta$$

$$= \frac{1}{2\pi i} \int_\gamma \frac{f(\zeta) \, d\zeta}{(\zeta - z_0 - h)(\zeta - z_0)}$$

Observe that

$$\frac{1}{(\zeta - z_0 - h)(\zeta - z_0)} = \frac{1}{(\zeta - z_0)^2} + \frac{h}{(\zeta - z_0)^2(\zeta - z_0 - h)}$$

Fig. 4.4

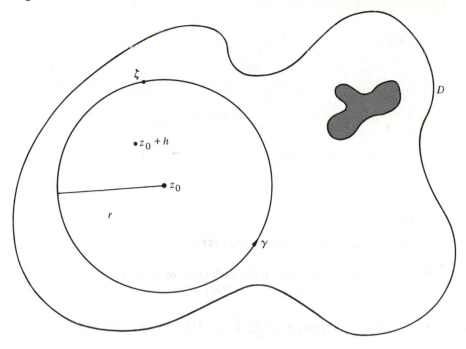

Hence

$$\frac{f(z_0 + h) - f(z_0)}{h} = \frac{1}{2\pi i} \int_\gamma \frac{f(\zeta)}{(\zeta - z_0)^2} \, d\zeta$$

$$+ \frac{h}{2\pi i} \int_\gamma \frac{f(\zeta) \, d\zeta}{(\zeta - z_0)^2(\zeta - z_0 - h)} \qquad (4.16)$$

By Exercise 2.5.4 there exists a constant M such that $|f(\zeta)| \leq M$ for $\zeta \in \gamma$. Let $|h| < r/2$. Then if $\zeta \in \gamma$,

$$|\zeta - z_0 - h| \geq |\zeta - z_0| - |h| > r - \frac{r}{2} = \frac{r}{2}$$

Thus from Eq. (4.16) we obtain

$$\left| \frac{f(z_0 + h) - f(z_0)}{h} - \frac{1}{2\pi i} \int_\gamma \frac{f(\zeta)}{(\zeta - z_0)^2} \, d\zeta \right| \leq \frac{|h|}{2\pi} \frac{M}{r^2(r/2)} \, 2\pi r = \frac{2M}{r^2} |h|$$

which approaches zero as h approaches zero. This proves that

$$f'(z_0) = \frac{1}{2\pi i} \int_\gamma \frac{f(\zeta)}{(\zeta - z_0)^2} \, d\zeta \qquad (4.17)$$

By using (4.17), we shall show $f''(z_0) = [(2!)/(2\pi i)] \int_\gamma f(\zeta)/(\zeta - z_0)^3 \, d\zeta$. In fact, by using (4.17) and the notation of this theorem, we have

$$\left| \frac{f'(z_0 + h) - f'(z_0)}{h} - \frac{2!}{2\pi i} \int_\gamma \frac{f(\zeta)}{(\zeta - z_0)^3} \, d\zeta \right|$$

$$= \left| \frac{1}{2\pi i} \int_\gamma \frac{f(\zeta)}{h} \left[\frac{1}{(\zeta - z_0 - h)^2} - \frac{1}{(\zeta - z_0)^2} - \frac{2h}{(\zeta - z_0)^3} \right] d\zeta \right|$$

$$= \frac{1}{2\pi} \left| \int_\gamma \frac{f(\zeta)}{h} \frac{(\zeta - z_0)^3 - (\zeta - z_0)(\zeta - z_0 - h)^2 - 2h(\zeta - z_0 - h)^2}{(\zeta - z_0)^3(\zeta - z_0 - h)^2} \, d\zeta \right|$$

$$= \frac{1}{2\pi} \left| \int_\gamma \frac{f(\zeta)}{h} \frac{3(\zeta - z_0)h^2 - 2h^3}{(\zeta - z_0)^3(\zeta - z_0 - h)^2} \, d\zeta \right|$$

$$\leq \frac{1}{2\pi} \frac{M}{r^3(r/2)^2} (3r|h| + 2|h|^2)$$

which approaches zero as h approaches zero. The proof is complete.

Corollary 4.1 Let f be analytic in the disk $|z - a| < R$. Let $0 < r < R$ and let $\gamma(t) = a + re^{2\pi it}$ for $0 \leq t \leq 1$. Then if $|z_0 - a| < r$ and $n \geq 0$ is an integer,

$$f^{(n)}(z_0) = \frac{n!}{2\pi i} \int_\gamma \frac{f(\zeta)}{(\zeta - z_0)^{n+1}} \, d\zeta \qquad (4.18)$$

Proof. The proof is by induction on n. Theorem 4.4 gives us Eq. (4.18) for $n = 0$. So suppose $n \geq 0$ is an integer and (4.18) holds for n and for all functions analytic on the disk $|z - a| < R$. We shall show that (4.18) holds for $n + 1$ and for all functions analytic on the disk $|z - a| < R$. In fact,

$$f^{(n+1)}(z_0) = (f')^{(n)}(z_0) = \frac{n!}{2\pi i} \int_\gamma \frac{f'(\zeta)}{(\zeta - z_0)^{n+1}} \, d\zeta$$

$$= \frac{(n + 1)!}{2\pi i} \int_\gamma \frac{f(\zeta)}{(\zeta - z_0)^{n+2}} \, d\zeta$$

where the last equality is accomplished by integrating by parts (see Exercise 4.2.4).

Corollary 4.2 If $f(z) = u(x,y) + iv(x,y)$ is analytic in a domain D, then all partial derivatives of u and v exist and are continuous in D.

Proof. Let f be analytic in D. Then as f' is also analytic in D, it follows from formulas (3.7) and (3.8) that

$$f' = u_x + iv_x = v_y - iu_y \qquad (4.19)$$

Since f' is analytic, it is certainly continuous, and therefore the four partial derivatives u_x, v_x, u_y, and v_y are continuous, as they are the real and imaginary parts of a continuous function. The analyticity of f' and the formulas (3.7) and (3.8) applied to Eq. (4.19) imply that

$$f'' = u_{xx} + iv_{xx} \qquad f'' = v_{yx} - iu_{yx}$$

$$f'' = v_{xy} - iu_{xy} \qquad f'' = -u_{yy} - iv_{yy}$$

Since f'' is analytic, it is certainly continuous, and so the eight partial derivatives u_{xx}, u_{yy}, u_{xy}, u_{yx}, v_{xx}, v_{yy}, v_{xy}, and v_{yx} are also continuous, since they are the real and imaginary parts of a continuous function. The result of the corollary follows by induction. (See Exercise 4.4.9.)

Example 4.6 Show that the following identities hold.

a. $\displaystyle\int_\gamma \frac{e^z + \sin z}{z}\, dz = 2\pi i$, if γ is the circle $|z| = 1$.

b. $\displaystyle\int_\gamma \frac{\sin z}{(z - \pi)^3}\, dz = 0$, if γ is the circle $|z| = 4$.

c. $\displaystyle\int_\gamma \frac{dz}{(z - 1)(z + 3)} = \frac{\pi i}{2}$, if γ is the circle $|z| = 2$.

d. $\displaystyle\int_\gamma \frac{dz}{z(z - 1)} = 0$, if γ is the circle $|z| = 3$.

Proof
a. By using formula (4.12) with $f(z) = e^z + \sin z$ and $z_0 = 0$, we get

$$\int_\gamma \frac{e^z + \sin z}{z}\, dz = 2\pi i f(0) = 2\pi i$$

b. By using formula (4.18) with $f(z) = \sin z$, $z_0 = \pi$, and $n = 2$, we get

$$\int_\gamma \frac{\sin z}{(z - \pi)^3}\, dz = \pi i f''(\pi) = 0$$

c. First we have

$$\frac{1}{(z - 1)(z + 3)} = \frac{1/4}{z - 1} - \frac{1/4}{z + 3}$$

Hence

$$\int_\gamma \frac{dz}{(z - 1)(z + 3)} = \frac{1}{4}\int_\gamma \frac{dz}{z - 1} - \frac{1}{4}\int_\gamma \frac{dz}{z + 3}$$

$$= \frac{1}{4}\, 2\pi i - 0 = \frac{\pi i}{2}$$

where the last integral is zero by Theorem 4.3 since $1/(z + 3)$ is analytic inside the star domain $|z| < \frac{7}{2}$ and γ is a closed contour lying in D.

d. Note that

$$\frac{1}{z(z-1)^2} = \frac{1}{z} - \frac{1}{z-1} + \frac{1}{(z-1)^2}$$

and so

$$\int_\gamma \frac{dz}{z(z-1)^2} = \int_\gamma \frac{1}{z}\, dz - \int_\gamma \frac{1}{z-1}\, dz + \int_\gamma \frac{1}{(z-1)^2}\, dz$$

$$= 2\pi i - 2\pi i + 0 = 0$$

Theorem 4.6 Let f be analytic in the disk $|z - a| < R$ and assume that $|f(z)| \le M$ in this disk. Then the following *Cauchy's inequality* holds:

$$|f^{(n)}(a)| \le \frac{Mn!}{R^n} \tag{4.20}$$

Proof. By using Eq. (4.18) with γ being the circle $|z - a| = r < R$ and $z_0 = a$, we have

$$|f^{(n)}(a)| \le \frac{n!}{2\pi} \frac{M}{r^{n+1}} 2\pi r = \frac{Mn!}{r^n}$$

Since this holds for all $r < R$, estimate (4.20) follows.

The following theorem is called Liouville's theorem (Joseph Liouville, 1809–1882, French), though it was first proved by Cauchy.

JOSEPH LIOUVILLE

Joseph Liouville was born into a distinguished family in St. Omer, France, in 1809. He studied at the École Polytechnique and was appointed professor of mathematics there in 1833. In 1836 he founded the *Journal des Mathématiques Pures et Appliquées*, which upheld the high standard of French mathematics throughout the 19th century. He was appointed professor at the Sorbonne and the Collège de France in 1839.

Liouville contributed significantly to many fields of mathematics, especially to boundary-value problems for second-order linear differential equations. He was also highly interested in number theory, a topic on which he published over 200 papers. Today he is also remembered for having published the works of Galois, after the latter's untimely death.

Liouville lived to be 73 years old; he died in Paris on September 8, 1882.

Theorem 4.7. *Liouville's Theorem*   A bounded entire function is identically constant.

Proof. Let $|f(z)| \leq M$ for all $z \in \mathbf{C}$. Then by Eq. (4.20) with $n = 1$, any arbitrary point a, and any radius R, we get

$$|f'(a)| \leq \frac{M}{R} \to 0 \qquad \text{as } R \to \infty$$

Hence $f'(z) = 0$ for all $z \in \mathbf{C}$ and therefore f is a constant function.

The following result is due to Gauss.

CARL FRIEDRICH GAUSS

Carl Friedrich Gauss, one of the greatest mathematicians of all time, was born in a cottage at Brunswick, Germany, on April 30, 1777. His family was very poor. His father, a gardener and bricklayer, was a rough man with stubborn views who tried to prevent the education of his son. Carl never felt any real affection for his father. His mother, however, was always on his side and encouraged the education of her son.

Gauss was an infant prodigy. At the age of three he discovered a mistake that his father had made on the weekly payroll for his workers. When he was 10, he astonished his teacher when in a few seconds he computed the sum $1 + 2 + 3 \cdots + 100$ [undoubtedly through the formula $100(100 + 1)/2$, which gives the sum of this arithmetic progression]. It is said that he conceived nearly all his fundamental mathematical ideas between the ages of 14 and 17.

The education of the poor boy (and later the marriages of the poor scientist) became possible through the financial assistance of Carl Wilhelm Ferdinand, Duke of Brunswick. Gauss studied at the University of Göttingen and received his doctorate from the University of Helmstäd in 1798. In his doctoral thesis he proved the fundamental theorem of algebra (see Theorem 4.8).

Gauss was married on October 9, 1805, to Johanne Osthof. Three children were born of this marriage. Johanne died on October 11, 1809, and for the sake of his young children he married again on August 4, 1810. Three children were also born of this marriage.

(Continued)

Theorem 4.8. *Fundamental Theorem of Algebra*   A polynomial P of degree $n \geq 1$ has at least one root.

Proof. The proof is by contradiction. Let

$$P(z) = a_n z^n + a_{n-1} z^{n-1} + \cdots + a_1 z + a_0 \qquad a_n \neq 0$$

be a polynomial of degree n for which the theorem is false. Then $P(z) \neq 0$ for all $z \in \mathbf{C}$, and therefore the function F defined by $F(z) = 1/P(z)$ is an entire function. We shall now prove that F is bounded on $\mathbf{C}$. To prove this, observe that

$$P(z) = z^n \left(a_n + \frac{a_{n-1}}{z} + \cdots + \frac{a_1}{z^{n-1}} + \frac{a_0}{z^n} \right)$$

For $|z|$ sufficiently large, clearly

$$\left| \frac{a_{n-1}}{z} + \cdots + \frac{a_1}{z^{n-1}} + \frac{a_0}{z^n} \right| < \frac{|a_n|}{2}$$

and hence

$$\left| a_n + \frac{a_{n-1}}{z} + \cdots + \frac{a_1}{z^{n-1}} + \frac{a_0}{z^n} \right| \geq |a_n| - \left| \frac{a_{n-1}}{z} + \cdots + \frac{a_0}{z^n} \right|$$

$$> |a_n| - \frac{|a_n|}{2} = \frac{|a_n|}{2}$$

(Continued)

From 1807 to the end of his life, Gauss was professor of mathematics and director of the observatory at the University of Göttingen. For his many ingenious discoveries he has been called the "prince of mathematicians." He published 155 papers during his life, but most of his discoveries became known only posthumously from his diaries and correspondence. Gauss made outstanding contributions to celestial mechanics, magnetism, electricity, mathematical physics, algebra, analysis, differential geometry, and other fields.

He was often criticized for not expressing any praise or enthusiasm for the discoveries of his contemporaries, whom he seemed to ignore, but his diaries and correspondence may help justify his peculiar behavior. Gauss himself had made the same discoveries long before the others and he simply kept quiet without claiming any priority.

Gauss was the recipient of many awards and honors during his later life. He died early in the morning of February 23, 1855, at the age of 77.

Thus there exists $r > 0$ such that if $|z| \geq r$,

$$|F(z)| = \frac{1}{|P(z)|} < \frac{2}{|a_n| \, |z|^n}$$

Let $k = 2/(|a_n|r^n)$. In the compact set $|z| \leq r$, $|F(z)|$ is bounded by a constant L. If $M = \max(k, L)$, it follows that

$$|F(z)| \leq M \qquad \text{for all } z \in \mathbf{C}$$

Thus F is a bounded entire function. By Liouville's theorem, $F(z) = $ constant Hence P is constant and this contradiction proves the result.

Finally let us prove Morera's theorem (Giacinto Morera, 1856–1909, Italian), which is a form of a converse of Cauchy's integral theorem 4.3.

Theorem 4.9. *Morera's Theorem* Let f be a continuous function in a domain D and let

$$\int_\gamma f(z)\, dz = 0 \qquad\qquad (4.21)$$

for every triangle γ which together with its interior is in D. Then f is analytic in D.

Proof. As we mentioned in the proof of Lemma 4.1, the analyticity of f was not required to get the conclusion there. Only Eq. (4.9) was required, and this is our assumption (4.21) in this theorem. Therefore, we have the conclusion of Lemma 4.1 for any disk d contained in D. That is, there is an analytic function F on d such that

$$F'(z) = f(z) \qquad \text{for } z \in d$$

Since F is analytic on d, it follows by Theorem 4.5 that F' is also analytic on d. Since d is arbitrary, the proof is complete.

Example 4.7 If $M \geq 0$ and f is an entire function such that $\operatorname{Re} f(z) \leq M$ for all $z \in \mathbf{C}$, then f is constant.

Proof. Consider the function

$$F(z) = e^{f(z)} \qquad \text{for } z \in \mathbf{C}$$

Then F is an entire function and $|F(z)| = e^{\operatorname{Re} f(z)} \leq e^M$. By Liouville's theorem, F is constant, and therefore f (which is a continuous function) must be constant also, because it is discrete-valued (see Exercise 2.6.1).

EXERCISES

4.4.1 Show that

$$\int_\gamma \frac{\cos z}{z(z^2 + 1)} \, dz = \begin{cases} 2\pi i(1 - \cos i) & \gamma: |z| = 3 \\ 2\pi i & \gamma: |z| = 1/3 \\ 0 & \gamma: |z - 1| = 1/3 \end{cases}$$

4.4.2 Show that

$$\int_\gamma \frac{\cos z}{z^2(z - 1)} \, dz = \begin{cases} 0 & \gamma: |z + 1| = 1/2 \\ 2\pi i(-1 + \cos 1) & \gamma: |z| = 3/2 \\ 2\pi i \cos 1 & \gamma: |z - 1| = 1/2 \end{cases}$$

4.4.3 Evaluate the following integrals.

a. $\displaystyle\int_\gamma \frac{z^3 - 3z + 1}{(z - i)^2} \, dz$ on $\gamma: |z| = 2$

b. $\displaystyle\int_\gamma \frac{\cos z}{z^7} \, dz$ on $\gamma: |z - 1| = 2$

c. $\displaystyle\int_\gamma \frac{dz}{z^4 - 1}$ on $\gamma: |z| = 3$

4.4.4 Let $M \geq 0$ and let f be an entire function such that $\operatorname{Im} f(z) \leq M$ for all $z \in \mathbf{C}$. Show that f is constant.

4.4.5 Let $\gamma(t) = 2e^{2\pi i t} + 1 + i$, for $0 \leq t \leq 1$. Evaluate the integrals.

a. $\displaystyle\int_\gamma \frac{\sin z}{z^2 + 1} \, dz$ b. $\displaystyle\int_\gamma \frac{\cos z}{z^3 - z} \, dz$ c. $\displaystyle\int_\gamma \frac{e^z}{z^3} \, dz$

4.4.6 Let f be analytic in the disk $|z| \leq R$ and let $|f(z)| \leq M$ for all $|z| \leq R$. Prove that

$$|f^{(n)}(z)| \leq \frac{MRn!}{(R - |z|)^{n+1}} \qquad |z| < R$$

4.4.7 Let f be analytic in $|z - z_0| < R$. Prove Gauss's mean value theorem

$$f(z_0) = \frac{1}{2\pi} \int_0^{2\pi} f(z_0 + re^{i\theta}) \, d\theta$$

for any $0 < r < R$.

4.4.8 Show that

$$\int_0^{2\pi} e^{\cos \theta} \cos (\sin \theta) \, d\theta = 2\pi$$

4.4.9 Prove the inductive assertion that occurs in the proof of Corollary 4.2 by noting that

$$f^{(n)} = \frac{\partial^n u}{\partial x^n} + i \frac{\partial^n v}{\partial x^n}$$

and then proving by induction [with the aid of Eq. (4.19)] that if

$i_1, \ldots, i_m$ and $j_1, \ldots, j_m$ are nonnegative integers satisfying $i_1 + j_1 + \cdots + i_m + j_m = n$, then

$$\frac{\partial^n u}{\partial x^n} = (-1)^k \frac{\partial^n v}{\partial x^{i_1} \partial y^{j_1} \cdots \partial x^{i_m} \partial y^{j_m}} \qquad \begin{array}{l} \text{if } j_1 + \cdots + j_m = 2k + 1 \\ \text{for some } k \in \mathbf{Z} \end{array}$$

$$= (-1)^k \frac{\partial^n u}{\partial x^{i_1} \partial y^{j_1} \cdots \partial x^{i_m} \partial y^{j_m}} \qquad \begin{array}{l} \text{if } j_1 + \cdots + j_m = 2k \\ \text{for some } k \in \mathbf{Z} \end{array}$$

and

$$\frac{\partial^n v}{\partial x^n} = (-1)^k \frac{\partial^n u}{\partial x^{i_1} \partial y^{j_1} \cdots \partial x^{i_m} \partial y^{j_m}} \qquad \begin{array}{l} \text{if } j_1 + \cdots + j_m = 2k - 1 \\ \text{for some } k \in \mathbf{Z} \end{array}$$

$$= (-1)^k \frac{\partial^n v}{\partial x^{i_1} \partial y^{j_1} \cdots \partial x^{i_m} \partial y^{j_m}} \qquad \begin{array}{l} \text{if } j_1 + \cdots + j_m = 2k \\ \text{for some } k \in \mathbf{Z} \end{array}$$

4.5

THE TAYLOR-SERIES EXPANSION OF AN ANALYTIC FUNCTION

In Sec. 3.3 we proved that a power series represents an analytic function. Here we shall prove that a version of the converse is also true. In fact, we have the following theorem.

Theorem 4.10 Let $R > 0$, $z_0 \in \mathbf{C}$, and let f: $\{z: |z - z_0| < R\} \to \mathbf{C}$ be analytic. Then

$$f(z) = \sum_{n=0}^{\infty} \frac{f^{(n)}(z_0)}{n!} (z - z_0)^n \qquad \text{whenever } |z - z_0| < R \qquad (4.22)$$

That is, the power series converges and has f as its sum function.

Proof. Let $\gamma(t) = z_0 + Re^{it}$ for $0 \le t \le 2\pi$, and let $|z - z_0| < R$. Choose positive numbers r and r_1 such that $|z - z_0| \le r < r_1 < R$. Let Γ_1 and Γ be the circles centered at z_0, with radii r_1 and r, respectively. See Fig. 4.5. For any $\zeta \in \Gamma_1$ we have

$$|\zeta - z| = |(\zeta - z_0) - (z - z_0)| > |\zeta - z_0| - |z - z_0| \ge r_1 - r \qquad (4.23)$$

By the Cauchy integral formula, we have

$$f(z) = \frac{1}{2\pi i} \int_{\Gamma_1} \frac{f(\zeta)}{\zeta - z} \, d\zeta \qquad (4.24)$$

Observe that for any positive integer n and complex number $w \ne 1$,

$$\frac{1}{1 - w} = 1 + w + \cdots + w^{n+1} + \frac{w^n}{1 - w}$$

Fig. 4.5

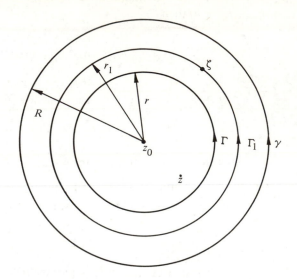

Thus

$$\frac{1}{\zeta - z} = \frac{1}{(\zeta - z_0) - (z - z_0)} = \frac{1}{\zeta - z_0}\frac{1}{1 - (z - z_0)/(\zeta - z_0)}$$

$$= \frac{1}{\zeta - z_0}\left\{1 + \left(\frac{z - z_0}{\zeta - z_0}\right) + \cdots + \left(\frac{z - z_0}{\zeta - z_0}\right)^{n-1}\right.$$

$$\left. + \frac{[(z - z_0)/(\zeta - z_0)]^n}{1 - (z - z_0)/(\zeta - z_0)}\right\}$$

$$= \frac{1}{\zeta - z_0} + \frac{(z - z_0)}{(\zeta - z_0)^2} + \cdots + \frac{(z - z_0)^{n-1}}{(\zeta - z_0)^n} + \frac{(z - z_0)^n}{(\zeta - z)(\zeta - z_0)^n}$$

(4.25)

Multiplying both sides of Eq. (4.25) by $(1/2\pi i)f(\zeta)$ and integrating along Γ_1, we obtain, in view of Eq. (4.24),

$$f(z) = f(z_0) + \frac{f'(z_0)}{1!}(z - z_0) + \frac{f''(z_0)}{2!}(z - z_0)^2 + \cdots$$

$$+ \frac{f^{(n-1)}(z_0)}{(n-1)!}(z - z_0)^{n-1} + R_n(z)$$

where

$$R_n(z) = \frac{(z - z_0)^n}{2\pi i}\int_{\Gamma_1}\frac{f(\zeta)\,d\zeta}{(\zeta - z)(\zeta - z_0)^n} \qquad (4.26)$$

Let M be the maximum value of $|f(\zeta)|$ on Γ_1. Then, using Eq. (4.23), we have

$$|R_n(z)| \le \frac{|z - z_0|^n}{2\pi} \left| \int_{\Gamma_1} \frac{f(\zeta)\, d\zeta}{(\zeta - z)(\zeta - z_0)^n} \right|$$

$$\le \frac{r^n}{2\pi} \frac{M}{(r_1 - r)r_1^n} 2\pi r_1 = \frac{M}{1 - r/r_1} \left(\frac{r}{r_1}\right)^n$$

Since $0 < r/r_1 < 1$, $|R_n(z)|$ approaches zero as $n \to \infty$.

The following corollary states the result (premised in Sec. 3.4) that every analytic function is locally a power series.

Corollary 4.3 Let f be analytic at z_0. Then there exists a unique sequence of complex numbers $a_0, a_1, a_2, \ldots$ and $\delta > 0$ such that if $|z - z_0| < \delta$, then

$$f(z) = \sum_{n=0}^{\infty} a_n(z - z_0)^n$$

In fact, if n is a nonnegative integer and $0 < r < \delta$, then a_n is given by

$$a_n = \frac{f^{(n)}(z_0)}{n!} = \frac{1}{2\pi i} \int_{|z-z_0|=r} \frac{f(z)\, d\zeta}{(z - z_0)^{n+1}}$$

Corollary 4.4 states that a polynomial can be completely factorized.

Corollary 4.4 Let p be a polynomial of degree n. Then there exist unique complex numbers $\lambda, \mu_1, \mu_2, \ldots, \mu_n$ such that if $z \in \mathbf{C}$, then

$$p(z) = \lambda(z - \mu_1)(z - \mu_2) \cdots (z - \mu_n)$$

The power series (4.22) is called the *Taylor series* of the function f about z_0 (Brook Taylor, 1685–1731, English). The circle of convergence of this series

BROOK TAYLOR

Brook Taylor was born at Norton, England, on December 9, 1685. He was educated at St. John's College, Cambridge. In 1712, he discovered the theorem that today bears his name, and the result was published in 1715. Taylor wrote many memoirs on series, logarithms, and physics. He is also noted for his contributions to the development of the calculus.

Taylor was elected a fellow of the Royal Society in 1712, and later that year was a member of the committee deciding between the rival claims of Newton and Leibniz. He died in London on December 29, 1731, at the age of 46.

Fig. 4.6

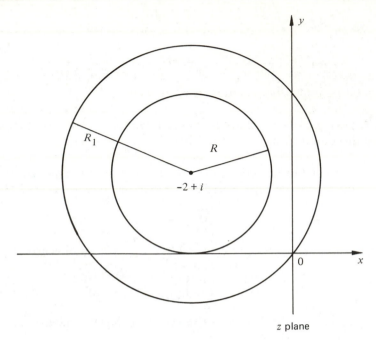

z plane

is at least as large as the largest circle centered at z_0 in which f is analytic. If we know that f is analytic in the interior of a circle γ of radius R, *no test is required* for the convergence of the Taylor series within γ. However, the series (4.22) may have a radius of convergence larger than R. The following example shows that this is possible.

Example 4.8 Find the Taylor-series expansion of the function $f(z) = \operatorname{Log} z$ about the point $z_0 = -2 + i$.

Solution. Since $\operatorname{Log} z$ is not continuous on the nonpositive real axis, the largest radius for which Theorem 4.10 can hold is $R = 1$. (See Fig. 4.6.) Note that $f(z_0) = \operatorname{Log}(-2 + i) = \operatorname{Log} \sqrt{5} + i \operatorname{Arg}(-2 + i)$. Furthermore, for all $z \in \mathbf{C} - \{x: x \le 0\}$, we have

$$f^{(n)}(z) = \frac{(-1)^{n-1}(n-1)!}{z^n} \qquad \text{for } n = 1, 2, \ldots$$

and therefore

$$f^{(n)}(z_0) = \frac{(-1)^{n-1}(n-1)!}{(-2+i)^n} = \frac{(-1)^{n-1}(n-1)!\,(2+i)^n}{(-5)^n}$$

$$= -\frac{(n-1)!\,(2+i)^n}{5^n}$$

Hence

$$\text{Log } z = {}^1\!/_2 \text{ Log } 5 + i \text{ Arg } (-2 + i)$$

$$-\sum_{n=1}^{\infty} \frac{1}{n} \left(\frac{2+i}{5} \right)^n (z + 2 - i)^n \tag{4.27}$$

The right-hand side in (4.27) represents Log in the disk $|z + 2 - i| < 1$. However, the radius of convergence of the series in (4.27) is (by Lemma 3.2)

$$R_1 = \frac{5}{|2 + i|} = \sqrt{5} > 1 = R$$

Next we shall prove a result, the corollary of which provides the motivation which was promised in Sec. 3.4.

Theorem 4.11 Let A be a domain and let $f: A \to \mathbf{C}$ be analytic. Let $B = \{z: f(z) = 0\}$. Then either B has no limit points in A, or else $B = A$.

Proof. Assume that B has a limit point $\zeta \in A$. Since f is continuous, $\zeta \in B$. Let $D = \{z:$ there exists a curve in B joining ζ to $z\}$. Clearly $\zeta \in D \subset B$. So D is a nonempty arcwise connected subset of B. It suffices to show that $D = A$. So suppose that it does not. As ζ is a limit point of B, by Theorems 4.10 and 3.2, there exists $\delta > 0$ such that if $|z - \zeta| < \delta$, then $f(z) = 0$. So it is clear that since D is arcwise connected, Theorems 4.10 and 3.2 imply that D is open. Since A and D are domains, $\varnothing \neq D \subset A$, and $A \neq D$, we see that there exists $z_0 \in A - D$ such that z_0 is a limit point of D. Again by Theorems 4.10 and 3.2, there exists $\rho > 0$ such that if $|z - z_0| < \rho$, then $f(z) = 0$. Thus because z_0 is a limit point of D and D is connected, it follows easily that $z_0 \in D$. This contradiction proves the theorem.

Corollary 4.5 Let A be a domain, and let $f: A \to \mathbf{C}$ and $g: A \to \mathbf{C}$ be analytic. Let $B = \{z: f(z) = g(z)\}$. Then either B has no limit points in A, or else $B = A$.

Theorem 4.12 Let $f(z) = \sum_{n=0}^{\infty} a_n(z - z_0)^n$ be a power series with radius of convergence $0 < R < \infty$. Then there exists w with $|w - z_0| = R$ such that f cannot be extended to be analytic on any open set containing w.

Proof. Suppose the theorem is false. Then if $|w - z_0| = R$, there exists an open set O_w containing w and an analytic extension

$$F_w: \{z: |z - z_0| < R\} \cup O_w \to \mathbf{C}$$

of f, that is an analytic function F_w on $\{z: |z - z_0| < R\} \cup O_w$ which agrees with f on the disk $|z - z_0| < R$. So as the circle $|z - z_0| = R$ is compact, clearly there exist points $w_1, \ldots, w_k$ on the circle $|z - z_0| = R$, numbers $V_1 > 0, \ldots, V_k > 0$ such that $\{z: |z - z_0| = R\} \subset \bigcup_{n=1}^{k} N_{V_n}(w_n)$, and analytic extensions $F_n: \{z: |z - z_0| < R\} \cup N_{V_n}(w_n) \to \mathbf{C}$ of f for $n = 1, \ldots, k$. So by Theorem 4.11, it is clear that there exists an analytic extension

Fig. 4.7

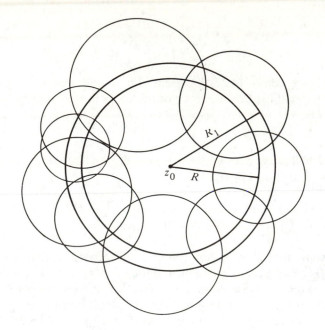

F: $\{z: |z - z_0| < R\} \cup \bigcup_{n=1}^{k} N_{V_n}(w_n) \to \mathbf{C}$ of f. Clearly there exists $0 < R < R_1 < \infty$ so that $\{z: |z - z_0| < R_1\} \subset \{z: |z - z_0| < R\} \cup \bigcup_{n=1}^{k} N_{V_n}(w_n)$. See Fig. 4.7.

If $|z - z_0| < R$, then $F(z) = f(z) = \sum_{n=0}^{\infty} a_n(z - z_0)^n$. Thus because F is analytic inside the circle $|z - z_0| < R_1$, it follows by Theorem 4.10 and Corollary 4.3 that if $|z - z_0| < R_1$, then $F(z) = \sum_{n=0}^{\infty} a_n(z - z_0)^n$. Hence R is not the radius of convergence of the power series $\sum_{n=0}^{\infty} a_n(z - z_0)^n$. This contradiction proves the theorem.

The point w in Theorem 4.12 is called a *nonremovable singularity* of the power series

$$\sum_{n=0}^{\infty} a_n(z - z_0)^n$$

Thus Theorem 4.12 is often stated as follows: "The radius of convergence of a power series is the distance from the center of convergence to the nearest nonremovable singularity of the power series."

The following result is an interesting consequence of Corollary 4.5 and Theorem 4.4.

Theorem 4.13. *Maximum Modulus Principle*   Let f be a nonconstant function analytic in a domain D. Then $|f|$ has no local maxima in D (and hence no absolute maxima in D either, since D is open).

Proof. If the theorem were false, there would exist a point $z_0 \in D$ and an ε neighborhood $N_\varepsilon(z_0)$ of z_0 whose closure is contained in D such that $|f(z_0)| \geq |f(z)|$ for all $z \in N_\varepsilon(z_0)$. Then by Theorem 4.4, for any positive $\delta \leq \varepsilon$, we have

$$
\begin{aligned}
f(z_0) &= \frac{1}{2\pi i} \int_{|z-z_0|=\delta} \frac{f(\zeta)}{\zeta - z_0} \, d\zeta \\
&= \frac{1}{2\pi i} \int_0^{2\pi} \frac{f(z_0 + \delta e^{i\theta})}{\delta e^{i\theta}} i\delta e^{i\theta} \, d\theta \\
&= \frac{1}{2\pi} \int_0^{2\pi} f(z_0 + \delta e^{i\theta}) \, d\theta
\end{aligned}
$$

We now claim that $|f(z_0 + \delta e^{i\theta}| = |f(z_0)|$ for $0 \leq \theta \leq 2\pi$. In fact,

$$|f(z_0 + \delta e^{i\theta})| \leq |f(z_0)| \qquad \text{for } 0 \leq \theta \leq 2\pi$$

and if strict inequality holds for some value of θ, it must hold (by the continuity of $|f(z)|$) on a segment of $[0,2\pi]$. In this case,

$$
\begin{aligned}
|f(z_0)| &\leq \frac{1}{2\pi} \int_0^{2\pi} |f(z_0 + \delta e^{i\theta})| \, d\theta \\
&< \frac{1}{2\pi} \int_0^{2\pi} |f(z_0)| \, d\theta \\
&= |f(z_0)|
\end{aligned}
$$

and this contradiction proves our claim. Since δ was any number in $(0,\varepsilon]$, the conclusion is that $|f(z)|$ is constant in $|z - z_0| \leq \varepsilon$. By the Cauchy-Riemann equations (or Exercise 3.6.4d), we see that $f(z) = f(z_0)$ for all z in $|z - z_0| \leq \varepsilon$. By Corollary 4.5 it now follows that $f(z) = f(z_0)$ for all $z \in D$, contradicting the fact that f is not constant in D. The proof is complete.

Corollary 4.5a If f satisfies the hypotheses of Theorem 4.13 and in addition $f(z) \neq 0$ for all $z \in D$, then $|f(z)|$ cannot have a local minimum in D (and hence no absolute minimum in D either).

Proof. Apply Theorem 4.13 to the function $1/f$.

The following corollary is a slightly different version of Cauchy's inequality.

Corollary 4.6 Let f be analytic on the disk $|z - a| < r$, and let $0 < R < r$. Then if $M = \sup_{|z-a|=R} |f(z)|$,

$$|f^n(a)| \leq \frac{Mn!}{R^n}$$

Proof. By Theorem 4.13,

$$\sup_{|z-a| \leq R} |f(z)| = \sup_{|z-a|=R} |f(z)|$$

and so the corollary follows by Theorem 4.6.

The following theorem, which yields a characterization of bounded analytic functions on the open unit disk, is due to Schwarz (1843–1921, German).

Theorem 4.14. *Schwarz's Lemma* Let f be analytic on the disk $|z| < 1$, let $f(0) = 0$, and let $|f(z)| \le M$ for all $|z| < 1$. Then $|f(z)| \le M|z|$ for $|z| < 1$.

Proof. There exist complex numbers $a_1, a_2, \ldots$ such that if $|z| < 1$, then $f(z) = \sum_{n=1}^{\infty} a_n z^n$. So if we let $g(z) = \sum_{n=0}^{\infty} a_{n+1} z^n$, then g is analytic on the disk $|z| < 1$, and moreover, $g(z) = f(z)/z$ if $0 < |z| < 1$. Thus if $0 < r < 1$, then $|z| = r$ implies

$$|g(z)| = \frac{|f(z)|}{|z|} \le \frac{M}{r}$$

Thus by Theorem 4.13, if $0 < r < 1$ and $|z| \le r$, then $|g(z)| \le M/r$. Since this is true for all $0 < r < 1$, we see that $|g(z)| \le M$ if $|z| < 1$. Hence $|f(z)| = |g(z)||z| \le M|z|$ if $|z| < 1$.

EXERCISES

4.5.1 Find the Taylor-series expansion of the following functions about the indicated points.

a. $\sin z$ about $z_0 = 0$
b. $\cos z$ about $z_0 = \pi$
c. $\sinh z$ about $z_0 = \pi/2$
d. $1/z$ about $z_0 = i$
e. $\dfrac{2z - 1}{z^2 - z}$ about $z_0 = i$
f. $z^3 - 4iz + 2$ about $z_0 = -2i$
g. $\text{Log}\,(1 - z)$ about $z_0 = i$

4.5.2 Let f be analytic in a bounded domain D and continuous on the boundary of D. Show that $|f|$ attains its maximum on the boundary of D.

HERMANN AMANDUS SCHWARZ

Hermann Amandus Schwarz, born in 1843, was a pupil of Weierstrass at the University of Berlin. He was a very original mathematician who contributed significantly to many branches of mathematics, including the theory of minimal surfaces and the theory of functions. In 1897 he succeeded Weierstrass as professor of mathematics at Berlin. He died in 1921.

4.5.3 Let f and g be analytic functions on $|z| \le 1$. Assume that $|f(z)| < M$ for $|z| = 1$ and that $f(z) = z^2 g(z)$ on $|z| \le \frac{1}{2}$. Show that $|f(z)| < M/4$ for $|z| \le \frac{1}{2}$.

4.5.4 Let $z_1, z_2, \ldots, z_n$ be n points in $\mathbf{C}$, and let D be a bounded domain in $\mathbf{C}$. Show that the product $|z - z_1||z - z_2| \cdots |z - z_n|$ attains its maximum at some point $z \in$ bd D. Investigate where the minimum of the above product is attained.

4.5.5 Let f be an entire function such that, for large z,

$$|f(z)| \le M|z|^n$$

where M is a constant and $n \ge 0$ is an integer. Show that f is a polynomial of degree at most n.

4.5.6 Show that for every positive integer n,

$$\sin \frac{\pi}{n} \sin \frac{2\pi}{n} \cdots \sin \frac{(n-1)\pi}{n} = \frac{n}{2^{n-1}}$$

4.6

CAUCHY'S THEOREM FOR SIMPLY CONNECTED DOMAINS

In this section we are going to extend our concept of integration to include integrating analytic functions over curves (which re not necessarily contours) and then use our expanded notion to arrive at our final version of *Cauchy's theorem*.

Before starting our discussion, we first recall the following remark from Sec. 2.3.

Remark 4.3 Let $A \subset \mathbf{C}$ and let $f: A \to \mathbf{C}$ be a function. Then f is continuous if and only if, for every open set $G_1 \subset \mathbf{C}$, there exists an open set $G_2 \subset \mathbf{C}$ such that $f^{-1}[G_1] = A \cap G_2$.

Let f be analytic at z_0. Then an analytic function $g: N \to \mathbf{C}$ is called a *primitive of f near z_0* if $z_0 \in N$ and $g'(z) = f(z)$ for all $z \in N$.

Lemma 4.2 Let $f: D \to \mathbf{C}$ be analytic, and let $\gamma: [a,b] \to D$ be a curve. Then there exists a curve $\Gamma: [a,b] \to \mathbf{C}$ such that if $a \le t \le b$, then there exists a primitive g of f near $\gamma(t)$ such that for all s near t with $a \le s \le b$, we have $\Gamma(s) = g(\gamma(s))$. Moreover, Γ is unique up to an additive constant.

Proof. We see by Corollaries 3.2 and 4.3 that for each $t \in [a,b]$ there exists a neighborhood N_t of $\gamma(t)$ and a primitive $g_t: N_t \to \mathbf{C}$ of f near $\gamma(t)$. So as γ is continuous, it follows from Remark 4.3 and Corollary 2.3 that there exist real numbers $a = t_0 < t_1 < \cdots < t_n = b$ such that for $1 \le k \le n$, there exists $s_k \in [a,b]$ such that $\gamma[[t_{k-1},t_k]] \subset N_{s_k}$. For $t_0 \le t \le t_1$, we define $\Gamma(t) = g_{s_1}(\gamma(t))$. For $t_1 \le t \le t_2$, we define $\Gamma(t) = g_{s_2}(\gamma(t)) - g_{s_2}(\gamma(t_1)) + g_{s_1}(\gamma(t_1))$. Continuing this process inductively, we obtain a function

$\Gamma: [a,b] \to \mathbf{C}$ which clearly has the desired property. To see that Γ is continuous, we need only note that on $N_{s_j} \cap N_{s_{j+1}}$, $g'_{s_j}(z) = g'_{s_{j+1}}(z)$, and so g_{s_j} and $g_{s_{j+1}}$ differ by a constant on $N_{s_j} \cap N_{s_{j+1}}$. This last remark also proves that Γ is unique up to an additive constant, and the proof of the lemma is complete.

The process of constructing Γ is called *analytically continuing the primitive of f along γ*, and Γ is called the *analytical continuation of the primitive of f along γ*. As Γ is unique up to an additive constant, the difference $\Gamma(b) - \Gamma(a)$ is independent of the choice of Γ and depends only on f and γ. Note that if γ is differentiable at t, then $\Gamma'(t) = f(\gamma(t))\gamma'(t)$, and so if γ is piecewise differentiable (that is, a contour), then

$$\int_\gamma f(z)\, dz = \int_a^b f(\gamma(t))\gamma'(t)\, dt$$

$$= \int_a^b \Gamma'(t)\, dt = \Gamma(b) - \Gamma(a)$$

Noting the above, we are now ready to extend our definition of the integral. If $f: D \to \mathbf{C}$ is analytic and $\gamma: [a,b] \to D$ is a curve, we let Γ be as in Lemma 4.2, and we define

$$\int_\gamma f(z)\, dz = \Gamma(b) - \Gamma(a)$$

The following two familiar properties still hold.

Lemma 4.3 Let $f: D \to \mathbf{C}$ be analytic and let $\gamma: [a,b] \to D$ be a curve. Then

a. $\displaystyle\int_\gamma f'(z)\, dz = f(\gamma(b)) - f(\gamma(a))$.

b. If $\gamma_1: [a,c] \to D$ and $\gamma_2: [c,b] \to D$ such that $\gamma = \gamma_1 + \gamma_2$, then

$$\int_\gamma f(z)\, dz = \int_{\gamma_1} f(z)\, dz + \int_{\gamma_2} f(z)\, dz$$

Proof
a. Since $\Gamma = f \circ \gamma$ is a solution of Lemma 4.2 for f' and γ,

$$\int_\gamma f'(z)\, dz = f(\gamma(b)) - f(\gamma(a))$$

b. Let Γ_1 and Γ_2 be as in Lemma 4.2 for γ_1 and γ_2, respectively. Then $\Gamma: [a,b] \to \mathbf{C}$ defined by

$$\Gamma(t) = \begin{cases} \Gamma_1(t) & \text{for } a \le t \le c \\ \Gamma_2(t) - \Gamma_2(c) + \Gamma_1(c) & \text{for } c \le t \le b \end{cases}$$

works for $\gamma = \gamma_1 + \gamma_2$, and so

$$\int_\gamma f(z)\,dz = \Gamma_2(b) - \Gamma_2(c) + \Gamma_1(c) - \Gamma_1(a)$$

$$= \int_{\gamma_2} f(z)\,dz + \int_{\gamma_1} f(z)\,dz$$

We are now ready to prove the main theorem of this section.

Theorem 4.15 Let $f\colon D \to \mathbf{C}$ be analytic, and let $H\colon [c,d] \times [a,b] \to D$ be continuous. Referring to Fig. 4.8, let

$$\begin{aligned}
\mu_0(s) &= H(s,a) && \text{for } c \le s \le d \\
\mu_1(s) &= H(s,b) && \text{for } c \le s \le d \\
\gamma_0(t) &= H(c,t) && \text{for } a \le t \le b \\
\gamma_1(t) &= H(d,t) && \text{for } a \le t \le b
\end{aligned}$$

Then

$$\int_{\mu_0} f(z)\,dz + \int_{\gamma_1} f(z)\,dz - \int_{\mu_1} f(z)\,dz - \int_{\gamma_0} f(z)\,dz = 0$$

Proof. For $(s,t) \in [c,d] \times [a,b]$, there exists a primitive $F_{(s,t)}\colon N_{(s,t)} \to \mathbf{C}$ of f near $H(s,t)$. So it follows from Remark 4.3 and Theorem 2.9 that there

Fig. 4.8

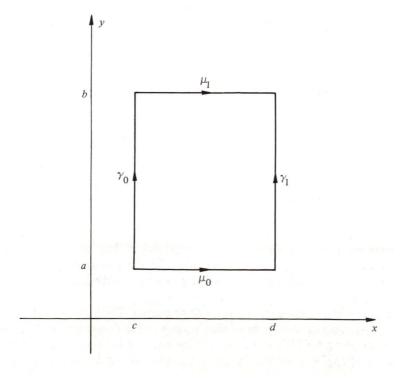

Fig. 4.9

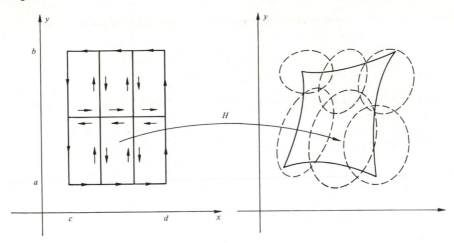

exist real numbers $c = s_0 < s_1 < \cdots < s_m = d$ and $a = t_0 < t_1 < \cdots < t_n = b$ such that if $1 \leq k \leq m$ and $1 \leq l \leq n$, then there exists $(s,t) \in [c,d] \times [a,b]$ such that $H[[s_{k-1}, s_k] \times [t_{l-1}, t_l]] \subset N_{(s,t)}$. For each (k,l) as above, let $R_{(k,l)}$ be the (oriented) boundary of $[s_{k-1}, s_k] \times [t_{l-1}, t_l]$. See Fig. 4.9.

Recall that for each rectangle $R_{(k,l)}$, there exists $(s,t) \in [c,d] \times [a,b]$ such that $H[R_{(k,l)}] \subset N_{(s,t)}$. Hence

$$\int_{R_{(k,l)}} f(z)\, dz = \int_{R_{(k,l)}} F'_{(k,l)}(z)\, dz = 0$$

Thus it follows that

$$\int_{\mu_0} f(z)\, dz + \int_{\gamma_1} f(z)\, dz - \int_{\mu_1} f(z)\, dz - \int_{\gamma_1} f(z)\, dz$$

$$= \sum_{k=1}^{m} \sum_{l=1}^{n} \int_{R_{(k,l)}} f(z)\, dz = 0$$

which concludes the proof.

Let $\gamma_0 : [a,b] \to D$ and $\gamma_1 : [a,b] \to D$ be curves, and let $I = [0,1]$. Then we say that γ_0 and γ_1 are *homotopic in D* if and only if there exists a continuous function $H : I \times [a,b] \to D$ such that if $t \in [a,b]$, then $H(0,t) = \gamma_0(t)$ and $H(1,t) = \gamma_1(t)$; in such a case, we call H a *homotopy* between γ_0 and γ_1 in D. See Fig. 4.10.

Let $\gamma_0 : [a,b] \to D$ and $\gamma_1 : [a,b] \to D$ be curves. Then we say that γ_0 and γ_1 are *homotopic in D relative to their endpoints* (or *homotopic in D keeping their endpoints fixed*) if there exists a homotopy H between γ_0 and γ_1 in D satisfying $H(t,a) = \gamma_0(a)$ and $H(t,b) = \gamma_0(b)$ for all $t \in I$. See Fig. 4.11.

Fig. 4.10

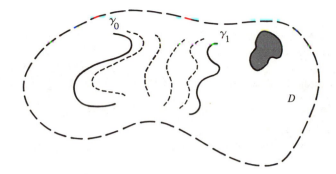

Note in particular that in the above definition, the curves γ_0 and γ_1 have the same endpoints. The following corollaries state that if we "deform a curve slightly" in the domain of an analytic function, the integral remains the same.

Corollary 4.7 Let $f: D \to \mathbf{C}$ be analytic; let $\gamma_0: [a,b] \to D$ and $\gamma_1: [a,b] \to D$ be two curves that are homotopic in D relative to their endpoints. Then

$$\int_{\gamma_0} f(z) \, dz = \int_{\gamma_1} f(z) \, dz$$

Fig. 4.11

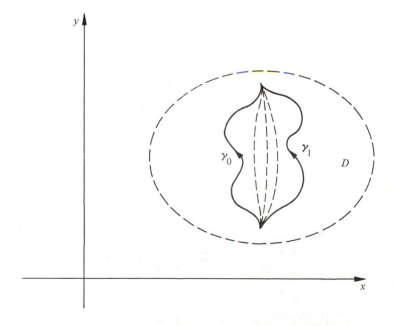

Proof. Applying theorem 4.15, we see that μ_0 and μ_1 are constants, and this implies the proof.

Let $\gamma_0: [a,b] \to D$ and $\gamma_1: [a,b] \to D$ be closed curves. Then we say that γ_0 and γ_1 are *homotopic in D through closed curves* if there exists a homotopy H between γ_0 and γ_1 in D satisfying $H(s,a) = H(s,b)$ for all $s \in I$ (in other words, for each $s \in I$, the curve $H_s: [a,b] \to D$ given by $H_s(t) = H(s,t)$ is closed). See Fig. 4.12.

Corollary 4.8 Let $f: D \to \mathbf{C}$ be analytic; let $\gamma_0: [a,b] \to D$ and $\gamma_1: [a,b] \to D$ be closed curves that are homotopic in D through closed curves. Then

$$\int_{\gamma_0} f(z) \, dz = \int_{\gamma_1} f(z) \, dz$$

Proof. Applying Theorem 4.15, we see that

$$\mu_0(s) = H(s,a) = H(s,b) = \mu_1(s)$$

for all $s \in I$, and so

$$\int_{\mu_0} f(z) \, dz = \int_{\mu_1} f(z) \, dz$$

which implies the corollary.

Fig. 4.12

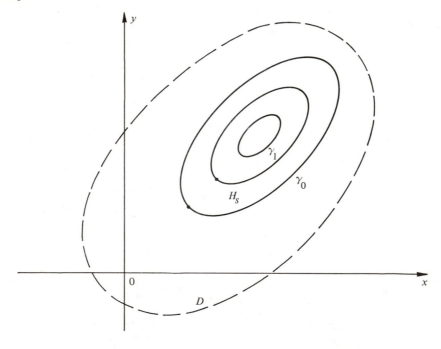

We say that a closed curve $\gamma \colon [a,b] \to D$ is *null-homotopic in D* or *contractible in D* if there exists a homotopy H in D through closed curves between γ and a constant curve. See Fig. 4.13.

Corollary 4.9 Let $f \colon D \to \mathbf{C}$ be analytic and let $\gamma \colon [a,b] \to D$ be a closed curve that is contractible in D. Then

$$\int_\gamma f(z)\, dz = 0$$

Proof. There exists a constant curve $\delta \colon [a,b] \to D$ to which γ is homotopic in D through closed curves. Then by Corollary 4.8, we have

$$\int_\gamma f(z)\, dz = \int_\delta f(z)\, dz = 0$$

We say that D is *simply connected* if it is connected and every closed curve in it is null-homotopic in D. Intuitively it means that D "has no holes."

Corollary 4.10 Let $f \colon D \to \mathbf{C}$ be analytic, and let D be simply connected.

Fig. 4.13

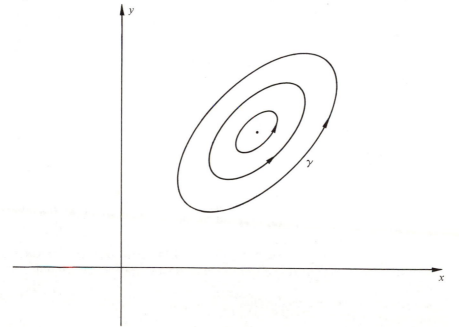

Then

a. If $\gamma: [a,b] \to D$ is a closed curve, then

$$\int_\gamma f(z)\, dz = 0$$

b. If $\gamma_0: [a,b] \to D$ and $\gamma_1: [a,b] \to D$ are curves such that $\gamma_0(a) = \gamma_1(a)$ and $\gamma_0(b) = \gamma_1(b)$, then

$$\int_{\gamma_0} f(z)\, dz = \int_{\gamma_1} f(z)\, dz$$

If f is an analytic function in a simply connected domain D, then Corollary 4.11 asserts that f has a primitive in D, while Corollary 4.12 states that if f is nowhere zero, then f "has a log."

Corollary 4.11 Let D be a simply connected domain and let $f: D \to \mathbf{C}$ be analytic. Then there exists an analytic function $F: D \to \mathbf{C}$ such that $F'(z) = f(z)$ for all $z \in D$.

Proof. Let $z_0 \in D$. We shall define $F: D \to \mathbf{C}$ as follows. If $z \in D$ and $\gamma: [a,b] \to D$ is a curve from z_0 to z, define $F(z) = \int_\gamma f(\zeta)\, d\zeta$. F is well defined (that is, F is a function) by Corollary 4.10, and it is analytic by Corollaries 4.3, 3.2, and Lemma 4.3.

Corollary 4.12 Let D be a simply connected domain, and let $f: D \to \mathbf{C} - \{0\}$ be analytic. Then there exists an analytic function $g: D \to \mathbf{C}$ such that $e^{g(z)} = f(z)$ for all $z \in D$.

Proof. Let $z_0 \in D$. By Corollary 4.11, it is clear that there exists an analytic function $g: D \to \mathbf{C}$ with $e^{g(z_0)} = f(z_0)$ such that $g'(z) = f'(z)/f(z)$ for all $z \in D$. Since

$$\left(\frac{e^g}{f}\right)'(z) = \frac{f(z)e^{g(z)}g'(z) - e^{g(z)}f'(z)}{[f(z)]^2} = 0$$

the proof is finished.

Example 4.9 Prove that

$$\int_0^\infty \frac{\sin x}{x}\, dx = \frac{\pi}{2}$$

Proof. Consider the closed contour γ of Fig. 4.14 oriented in the indicated direction. Clearly the function e^{iz}/z is analytic on a simply connected domain containing γ.

By Cauchy's theorem we thus have

$$\int_\gamma \frac{e^{iz}}{z}\, dz = 0$$

Fig. 4.14

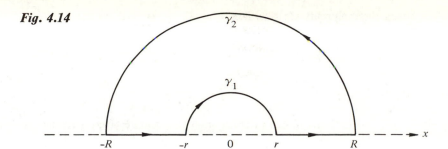

Observe that

$$\int_\gamma \frac{e^{iz}}{z}\, dz = \int_{-R}^{-r} \frac{e^{ix}}{x}\, dx + \int_{\gamma_1} \frac{e^{iz}}{z}\, dz + \int_r^R \frac{e^{ix}}{x}\, dx + \int_{\gamma_2} \frac{e^{iz}}{z}\, dz$$

Since $\tilde\gamma_1(t) = re^{it}$ and $\gamma_2(t) = Re^{it}$ for $0 \le t \le \pi$, we obtain

$$0 = \int_{-R}^{-r} \frac{e^{ix}}{x}\, dx - \int_0^\pi \frac{e^{ire^{it}}}{re^{it}}\, rie^{it}\, dt$$

$$+ \int_r^R \frac{e^{ix}}{x}\, dx + \int_0^\pi \frac{e^{iRe^{it}}}{Re^{it}}\, Rie^{it}\, dt$$

$$= J_1 + J_2 + J_3 + J_4 \tag{4.28}$$

We have

$$J_1 = \int_{-R}^{-r} \frac{e^{ix}}{x}\, dx = \int_R^r \frac{e^{-ix}}{x}\, dx = -\int_r^R \frac{e^{-ix}}{x}\, dx$$

Thus

$$J_1 + J_3 = \int_r^R \frac{e^{ix} - e^{-ix}}{x}\, dx$$

$$= 2i \int_r^R \frac{\sin x}{x}\, dx$$

$$\to 2i \int_0^\infty \frac{\sin x}{x}\, dx$$

as $r \to 0+$ and $R \to \infty$. Also

$$J_2 = -i \int_0^\pi e^{ir(\cos t + i \sin t)}\, dt$$

$$= -i \int_0^\pi e^{r(-\sin t + i \cos t)}\, dt$$

$$\to -i\pi$$

as $r \to 0$. Clearly

$$J_4 = i \int_0^\pi e^{iR(\cos t + i \sin t)}\, dt = i \int_0^\pi e^{-R \sin t + iR \cos t}\, dt$$

Thus, uisng the inequality $\sin t \geq 2t/\pi$ when $0 \leq t \leq \pi/2$, we have

$$|J_4| \leq \int_0^\pi e^{-R \sin t}\, dt = 2 \int_0^{\pi/2} e^{-R \sin t}\, dt$$

$$\leq 2 \int_0^{\pi/2} e^{-2Rt/\pi}\, dt = -\frac{\pi}{R}(e^{-R} - 1) \to 0$$

as $R \to \infty$. Taking limits in Eq. (4.28) as $r \to 0+$ and $R \to \infty$ gives us

$$2i \int_0^\infty \frac{\sin x}{x}\, dx - i\pi = 0$$

which proves the desired result.

The following example shows that one can sometimes compute real integrals very easily by using homotopy theory.

Example 4.10 Show that

$$\int_0^{2\pi} \frac{dt}{4 \cos^2 t + 9 \sin^2 t} = \frac{\pi}{3}$$

Solution. Let γ be the ellipse defined by $\gamma(t) = 2 \cos t + 3i \sin t$ for $0 \leq t \leq 2\pi$. Then since γ is homotopic through closed curves in $\mathbf{C} - \{0\}$ to the circle $|z| = 1$, we see that

$$2\pi i = \int_{|z|=1} \frac{dz}{z} = \int_\gamma \frac{dz}{z} = \int_\gamma \frac{-2 \sin t + 3i \cos t}{2 \cos t + 3i \sin t}\, dt$$

$$= \int_0^{2\pi} \frac{-2 \sin t + 3i \cos t}{2 \cos t + 3i \sin t} \cdot \frac{2 \cos t - 3i \sin t}{2 \cos t - 3i \sin t}\, dt$$

$$= \int_0^{2\pi} \frac{5 \sin t \cos t}{4 \cos^2 t + 9 \sin^2 t}\, dt + i \int_0^{2\pi} \frac{6}{4 \cos^2 t + 9 \sin^2 t}\, dt$$

from which the result follows.

EXERCISES

4.6.1 Let $f: D \to \mathbf{C}$ and $g: D \to \mathbf{C}$ be analytic and let $\gamma: [a,b] \to D$ be a curve. Prove the following statements.
  a. $\int_\gamma [f(z) + g(z)]\, dz = \int_\gamma f(z)\, dz + \int_\gamma g(z)\, dz$.
  b. If $\lambda \in \mathbf{C}$, then $\int_\gamma \lambda f(z)\, dz = \lambda \int_\gamma f(z)\, dz$.
  c. $\int_\gamma f(z)g'(z)\, dz = f(\gamma(b))g(\gamma(b)) - f(\gamma(a))g(\gamma(a)) - \int_\gamma g(z)f'(z)\, dz$.
  d. If $h: B \to \mathbf{C}$ is analytic and $f[\gamma[a,b]] \subset B$, then

$$\int_{f \circ \gamma} h(z)\, dz = \int_\gamma h(f(z))f'(z)\, dz$$

4.6.2 Find a simple closed curve $\gamma\colon [0,2\pi] \to \mathbf{C}$ with the ellipse

$$\frac{x^2}{4} + \frac{y^2}{9} = 1$$

as its trace. By calculating the integral $\int_\gamma dz/z$ in two different ways, show that

$$\int_0^{2\pi} \frac{dt}{1 + 2\cos^2 t} = \frac{2\pi}{3}.$$

4.7

A GENERALIZED CAUCHY FORMULA

In this section, we obtain a formula similar to Cauchy's formula which applies to curves that are merely closed. It will first be necessary for us to develop an additional property of the logarithmic function.

Lemma 4.4 Let D be a domain in $\mathbf{C}$ and let $g\colon D \to \mathbf{C}$ such that $g'(z) = 1/z$ for all $z \in D$. Then there exists $\lambda \in \mathbf{C}$ such that $e^{g(z)+\lambda} = z$ for all $z \in D$.

Proof. As D is a domain, there exists $z_0 \in D$ and $\varepsilon > 0$ such that $N_\varepsilon(z_0) \subset [D \cap (\mathbf{C} - \{x\colon x \le 0\})]$. Note that for all $z \in N_\varepsilon(z_0)$ we have $(g - \operatorname{Log})'(z) = 1/z - 1/z = 0$, and hence there exists $\lambda \in \mathbf{C}$ such that $\operatorname{Log} z = g(z) + \lambda$ for all $z \in N_\varepsilon(z_0)$; and so by Corollary 4.5 it follows that $e^{g(z)+\lambda} = z$ for all $z \in D$.

Lemma 4.5 Let $z_0 \in \mathbf{C}$, let $\gamma\colon [a,b] \to \mathbf{C} - \{z_0\}$ be a curve, let $f(z) = 1/(z - z_0)$ for all $z \in \mathbf{C} - \{z_0\}$, and let Γ be as in Lemma 4.2. Then after possibly modifying Γ by an additive constant, we have the identity

$$e^{\Gamma(t)} = \gamma(t) - z_0$$

for all $t \in [a,b]$.

Proof. Recall from the proof of Lemma 4.2 that there exists a partition $a = t_0 < t_1 < \cdots < t_n = b$ and analytic functions $g_1\colon N_1 \to \mathbf{C}, g_2\colon N_2 \to \mathbf{C},$ $\ldots,$ and $g_n\colon N_n \to \mathbf{C}$ such that if $1 \le j \le n$, then $g_j'(z) = 1/(z - z_0)$ and $\Gamma(s) = g_j(\gamma(s))$ for all $s \in [t_{j-1}, t_j]$. Let $G\colon \{z - z_0\colon z \in N_1\} \to \mathbf{C}$ be defined by $G(z - z_0) = g_1(z)$ for all $z \in N_1$. Then $G'(z) = 1/z$ for all $z \in \operatorname{dmn} G$. So by Lemma 4.4, after possibly modifying G by an additive constant, we may assume $e^{G(z)} = 1/z$ for each $z \in \operatorname{dmn} G$, and hence that $\Gamma(s) = e^{g_1(\gamma(s))} = 1/[\gamma(s) - z_0]$ whenever $s \in [t_0, t_1]$.

Now $g_1(z) = g_2(z)$ if $z \in N_1 \cap N_2$, and so by Corollary 4.5 it follows that

$$\Gamma(s) = e^{g_2(\gamma(s))} = \frac{1}{\gamma(s) - z_0}$$

for all $s \in [t_1, t_2]$. The proof follows by induction.

We shall now generalize a concept first mentioned in Sec. 4.2.

Let $z_0 \in \mathbf{C}$ and let $\gamma: [a,b] \to \mathbf{C} - \{z_0\}$ be a closed curve. Then we define the *index of γ at z_0* (or the *winding number of γ with respect to z_0*) to be

$$I(\gamma, z_0) = \frac{1}{2\pi i} \int_\gamma \frac{d\zeta}{\zeta - z_0}$$

Corollary 4.13 Let $z_0 \in \mathbf{C}$ and let $\gamma: [a,b] \to \mathbf{C} - \{z_0\}$ be a closed curve. Then $I(\gamma, z_0)$ is an integer.

Proof. By Lemma 4.5, there exists $\Gamma: [a,b] \to \mathbf{C}$ with $e^{\Gamma(t)} = \gamma(t) - z_0$ such that $\int_\gamma d\zeta/(\zeta - z_0) = \Gamma(b) - \Gamma(a)$. Thus since $e^{\Gamma(b)} = \gamma(b) - z_0 = \gamma(a) - z_0 = e^{\Gamma(a)}$, the result is clear.

The following example shows that all possible indices occur.

Example 4.11 Let n be an integer and let $z_0 \in \mathbf{C}$. Let $r > 0$, and let $\gamma_0(t) = z_0 + re^{2n\pi it}$ for $0 \le t \le 1$. Then if $z \in \mathbf{C}$, it follows that $I(\gamma_0, z) = n$ if $|z - z_0| < r$, and $I(\gamma_0, z) = 0$ if $|z - z_0| > r$.

Proof. Suppose $|z - z_0| < r$. Let $\gamma_1: [a,b] \to \mathbf{C}$ be the curve defined by $\gamma_1(t) = z + re^{2n\pi it}$ for $t \in [0,1]$. Then $H: [0,1] \times [0,1] \to \mathbf{C}$ defined by

$$H(s,t) = (1 - s)z_0 + sz + re^{2n\pi it}$$

provides a homotopy from γ_0 to γ_1 in $\mathbf{C} - \{z\}$ through closed curves. Hence

$$I(\gamma_0, z) = \frac{1}{2\pi i} \int_{\gamma_0} \frac{d\zeta}{\zeta - z} = \frac{1}{2\pi i} \int_{\gamma_1} \frac{d\zeta}{\zeta - z}$$

$$= \frac{1}{2\pi i} \int_0^1 \frac{2n\pi i re^{2n\pi irt}}{e^{2n\pi irt}} \, dt = n$$

Now suppose $|z - z_0| > r$. Then $h: [0,1] \times [0,1] \to \mathbf{C}$ defined by $h(s,t) = (1 - s)re^{2n\pi it} + z_0$ shows that γ_0 is homotopic to the constant curve $\delta(t) = z_0$ for all $t \in [0,1]$ in $\mathbf{C} - \{z\}$ through closed curves. Hence

$$I(\gamma_0, z) = \frac{1}{2\pi i} \int_{\gamma_0} \frac{d\zeta}{\zeta - z} = \frac{1}{2\pi i} \int_\delta \frac{d\zeta}{\zeta - z} = 0$$

The following theorem shows that in a certain sense closed curves are determined by their indices.

Theorem 4.16 Let $z_0 \in \mathbf{C}$, $\gamma_0: [a,b] \to \mathbf{C} - \{z_0\}$, and $\gamma_1: [a,b] \to \mathbf{C} - \{z_0\}$ be closed curves. Then γ_0 is homotopic through closed curves to γ_1 in $\mathbf{C} - \{z_0\}$ if and only if $I(\gamma_0, z_0) = I(\gamma_1, z_0)$.

Proof. If γ_0 is homotopic through closed curves to γ_1 in $\mathbf{C} - \{z_0\}$, then by Corollary 4.8, we have $I(\gamma_0, z_0) = I(\gamma_1, z_0)$. So suppose $I(\gamma_0, z_0) = I(\gamma_1, z_0)$.

Then by Lemma 4.5, there exist curves $\Gamma_0: [a,b] \to \mathbf{C}$ and $\Gamma_1: [a,b] \to \mathbf{C}$ with $e^{\Gamma_0(t)} = \gamma_0(t) - z_0$ and $e^{\Gamma_1(t)} = \gamma_1(t) - z_0$ for $t \in [a,b]$ such that $\Gamma_0(b) - \Gamma_0(a) = \Gamma_1(b) - \Gamma_1(a)$. Hence $\Gamma_1(a) - \Gamma_0(a) = \Gamma_1(b) - \Gamma_0(b)$. Let $H: [0,1] \times [a,b] \to \mathbf{C}$ be defined by

$$H(s,t) = e^{(1-s)\Gamma_0(t)+s\Gamma_1(t)} + z_0$$

if $0 \le s \le 1$ and $a \le t \le b$. Then clearly H is continuous and if $a \le t \le b$, then $H(0,t) = e^{\Gamma_0(t)} + z_0 = \gamma_0(t)$ and $H(1,t) = e^{\Gamma_1(t)} + z_0 = \gamma_1(t)$. Note that if $0 \le s \le 1$, then

$$H(s,a) = e^{(1-s)\Gamma_0(a)+s\Gamma_1(a)} + z_0$$
$$= e^{\Gamma_0(a)+s[\Gamma_1(a)-\Gamma_0(a)]} + z_0$$
$$= e^{\Gamma_0(a)+s[\Gamma_1(b)-\Gamma_0(b)]} + z_0$$
$$= e^{\Gamma_0(b)+s[\Gamma_1(b)-\Gamma_0(b)]} + z_0$$
$$\left[\text{since } e^{\Gamma_0(a)} = \gamma_0(a) = \gamma_0(b) = e^{\Gamma_0(b)}\right]$$
$$= e^{(1-s)\Gamma_0(b)+s\Gamma_1(b)} + z_0 = H(s,b)$$

Finally, as $e^z \ne 0$ for all z, we have $H(s,t) \ne z_0$ for all $(s,t) \in [0,1] \times [a,b]$, and so H is a homotopy through closed curves from γ_0 to γ_1 in $\mathbf{C} - \{z_0\}$.

Corollary 4.14 Let $\gamma: [a,b] \to \mathbf{C}$ be a closed curve, and $H: [0,1] \times [a,b] \to \mathbf{C}$ be a homotopy between closed curves of γ and a constant curve. Then if $z_0 \notin \operatorname{im} H$, $I(\gamma,z_0) = 0$.

Proof. Let $\delta: [a,b] \to \mathbf{C}$ be the constant curve defined by $\delta(t) = H(1,t)$. Then H is a homotopy between γ and δ in $\mathbf{C} - \{z_0\}$ through closed curves, and hence $I(\gamma,z_0) = I(\delta,z_0) = 0$.

Theorem 4.17 Suppose that $\gamma: [a,b] \to \mathbf{C}$ is a closed curve, and let $\phi: [0,1] \to \mathbf{C} - \operatorname{trace} \gamma$ be a curve. Then for all r, $s \in [0,1]$, we have $I(\gamma,\phi(r)) = I(\gamma,\phi(s))$.

Proof. For each $s \in [0,1]$, let $\gamma_s: [a,b] \to \mathbf{C}$ be the curve defined by $\gamma_s(t) = \gamma(t) - \phi(s)$. Then if $s \in [0,1]$, clearly γ_s is a closed curve in $\mathbf{C} - \{0\}$, and

$$I(\gamma,\phi(s)) = \frac{1}{2\pi i} \int_\gamma \frac{d\zeta}{\zeta - \phi(s)} = \frac{1}{2\pi i} \int_{\gamma_s} \frac{d\zeta}{\zeta} = I(\gamma_s,0)$$

Now if r, $s \in [0,1]$, it is clear that γ_r is homotopic to γ_s in $\mathbf{C} - \{0\}$ through closed curves, and hence that $I(\gamma,\phi(r)) = I(\gamma_r,0) = I(\gamma_s,0) = I(\gamma,\phi(s))$.

Theorem 4.18 Let $\gamma: [a,b] \to \mathbf{C}$ be a closed curve, and let $M = \max_{a \le t \le b} |\gamma(t)|$. Then $\{z: I(\gamma,z) \ne 0\} \subset \{z: |z| < M\}$.

Proof. Let $z \in \mathbf{C} -$ trace γ such that $|z| \geq M$. Let $h: [0,1] \times [a,b] \to \mathbf{C}$ be defined by $h(s,t) = (1 - s)\gamma(t)$. Then clearly, h is a homotopy through closed curves in $\mathbf{C} - \{z\}$ from γ to the constant curve $\delta(t) = 0$ for all $a \leq t \leq b$. Hence $I(\gamma,z) = I(\delta,z) = 0$, which proves the theorem.

Theorem 4.19 Let $f: D \to \mathbf{C}$ be analytic and let $z_0 \in D$. Suppose $\gamma_0: [a,b] \to D - \{z_0\}$ and $\gamma_1: [a,b] \to D - \{z_0\}$ are closed curves which are homotopic through closed curves in D. Then

$$\int_{\gamma_0} \frac{f(\zeta)}{\zeta - z_0}\, d\zeta - \int_{\gamma_1} \frac{f(\zeta)}{\zeta - z_0}\, d\zeta = 2\pi i [I(\gamma_0,z_0) - I(\gamma_1,z_0)]f(z_0)$$

Proof. Let $g: D \to \mathbf{C}$ be defined by the formula

$$g(z) = \begin{cases} \dfrac{f(z) - f(z_0)}{z - z_0} & z \in D - \{z_0\} \\ f'(z_0) & z = z_0 \end{cases}$$

Then g is clearly analytic in D, and so

$$\int_{\gamma_0} \frac{f(\zeta) - f(z_0)}{\zeta - z_0}\, d\zeta = \int_{\gamma_0} g(\zeta)\, d\zeta$$

$$= \int_{\gamma_1} g(\zeta)\, d\zeta$$

$$= \int_{\gamma_1} \frac{f(\zeta) - f(z_0)}{\zeta - z_0}\, d\zeta$$

from which the result follows.

The following corollary is the generalized Cauchy integral formula.

Corollary 4.15 Let $f: D \to \mathbf{C}$ be analytic and $z_0 \in D$. Let $\gamma: [a,b] \to D - \{z_0\}$ be a closed curve that is contractible in D. Then

$$\int_{\gamma} \frac{f(\zeta)}{\zeta - z_0}\, d\zeta = 2\pi i I(\gamma,z_0)f(z_0)$$

The proof of the following corollary is exactly the same as the proof of Corollary 4.1.

Corollary 4.16 Let $f: D \to \mathbf{C}$ be analytic, and suppose that $z_0 \in D$. Let $\gamma: [a,b] \to D - \{z_0\}$ be a closed curve that is contractible in D and let n be a nonnegative integer. Then

$$\int_{\gamma} \frac{f(\zeta)}{(\zeta - z_0)^{n+1}}\, d\zeta = 2\pi i I(\gamma,z_0)\frac{f^{(n)}(z_0)}{n!}$$

EXERCISES

4.7.1 Let

$$\gamma(t) = \begin{cases} e^{2\pi it} & 0 \le t \le 1 \\ 2 + e^{\pi i(1-2t)} & 1 \le t \le 2 \\ 2 + e^{\pi i(1+2t)} & 2 \le t \le 3 \end{cases}$$

Find the index of γ at the points $1/2$, $2 + i/2$, and $-3 + i$.

4.7.2 Compute the integrals

$$\int_\gamma \frac{e^z}{z^3} \, dz \quad \text{and} \quad \int_\gamma \frac{e^{\sin z}}{(z - 2)^9} \, dz$$

where γ is the same as in Exercise 4.7.1.

Chapter 5 Residue Theory

INTRODUCTION

We begin this chapter with the Laurent-series expansion of a function analytic in an annulus. We then use it to classify isolated singularities of an analytic function. The Cauchy residue theorem established in Sec. 5.4 is the heart of this chapter. Several applications of Cauchy's residue theorem to the evaluation of real integrals are presented in Sec. 5.5. Here the student will appreciate the usefulness of complex-function theory. Finally, in Sec. 5.6 we study the number of zeros and poles of mero-morphic functions. In particular we prove Rouché's theorem, which is often useful in determining the number of zeros of algebraic equations.

5.2
THE LAURENT EXPANSION OF AN
ANALYTIC FUNCTION

A generalization of the Taylor-series expansion of an analytic function is the following result of Laurent (Pierre Alphonse Laurent, 1813–1854, French).

Theorem 5.1 Let $0 \leq r_1 < r_2 \leq \infty$, let $z_0 \in \mathbf{C}$, and let f be analytic on the annulus $r_1 < |z - z_0| < r_2$. Let $r_1 < r < r_2$ and let γ be the circle $|z - z_0| = r$ (oriented in the counterclockwise direction). Finally, for each $n \in \mathbf{Z}$, let

$$a_n = \frac{1}{2\pi i} \int_\gamma \frac{f(\zeta)}{(\zeta - z_0)^{n+1}} \, d\zeta \tag{5.1}$$

Then if $r_1 < |z - z_0| < r_2$,

$$f(z) = \sum_{n=0}^{\infty} a_n(z - z_0)^n + \sum_{n=1}^{\infty} a_{-n}(z - z_0)^{-n} \tag{5.2}$$

The convergence of the two series in Eq. (5.2) is actually absolute, and so we may write

$$f(z) = \sum_{-\infty < n < \infty} a_n(z - z_0)^n$$

Proof. Let $r_1 < |z - z_0| < r_2$. We must show that Eq. (5.2) holds. Let $r_1 < R_1 < |z - z_0| < R_2 < r_2$, and let γ_1 and γ_2 be the circles $|\zeta - z_0| = R_1$ and $|\zeta - z_0| = R_2$, respectively. After considering the two domains D_1 and D_2 of Fig. 5.1, we see that Corollary 4.8 and the Cauchy integral formula imply

$$f(z) = \frac{1}{2\pi i} \int_{\mathrm{bd}\,D_1} \frac{f(\zeta)}{\zeta - z} \, d\zeta$$

while Corollary 4.7 implies

$$0 = \int_{\mathrm{bd}\,D_2} \frac{f(\zeta)}{\zeta - z} \, d\zeta$$

Fig. 5.1

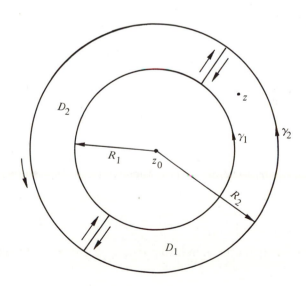

Then because the integrals along the straight-line segments cancel each other we have

$$f(z) = \frac{1}{2\pi i} \int_{\gamma_2} \frac{f(\zeta)}{\zeta - z}\, d\zeta - \frac{1}{2\pi i} \int_{\gamma_1} \frac{f(\zeta)}{\zeta - z}\, d\zeta$$

Following the method of the proof of Theorem 4.10 in exactly the same way, we obtain

$$\frac{1}{2\pi i} \int_{\gamma_2} \frac{f(\zeta)}{\zeta - z}\, d\zeta = \sum_{n=0}^{\infty} \left(\frac{1}{2\pi i} \int_{\gamma} \frac{f(\zeta)}{(\zeta - z_0)^{n+1}}\, d\zeta \right) (z - z_0)^n$$

After observing that

$$-\frac{1}{\zeta - z} = \frac{1}{z - \zeta} = \frac{1}{(z - z_0) - (\zeta - z_0)} = \frac{1}{z - z_0}\, \frac{1}{1 - (\zeta - z_0)/(z - z_0)}$$

$$= \frac{1}{z - z_0} + \frac{\zeta - z_0}{(z - z_0)^2} + \cdots + \frac{(\zeta - z_0)^{n-1}}{(z - z_0)^n} + \frac{(\zeta - z_0)^n}{(z - z_0)^n (z - \zeta)}$$

we see that an argument similar to that used in the proof of Theorem 4.10 shows that

$$-\frac{1}{2\pi i} \int_{\gamma_1} \frac{f(\zeta)}{\zeta - z}\, d\zeta = \sum_{n=1}^{\infty} \left[\frac{1}{2\pi i} \int_{\gamma_1} \frac{f(\zeta)}{(\zeta - z_0)^{-n+1}}\, d\zeta \right] (z - z_0)^{-n}$$

Thus since γ_1 and γ_2 are both homotopic in the annulus $r_1 < |\zeta - z_0| < r_2$ through closed curves to γ, we see that

$$f(z) = \sum_{n=0}^{\infty} a_n (z - z_0)^n + \sum_{n=1}^{\infty} a_{-n}(z - z_0)^{-n}$$

The series $f_1(z) = \sum_{n=0}^{\infty} a_n (z - z_0)^n$ is a power series in the annulus $r_1 < |z - z_0| < r_2$ and hence converges absolutely in the disk $|z - z_0| < r_2$. Since

$$\sum_{n=1}^{\infty} a_{-n}(\zeta - z_0)^{-n}$$

converges in the annulus $r_1 < |\zeta - z_0| < r_2$, we see that the power series

$$f_2(\zeta) = \sum_{n=1}^{\infty} a_{-n} \zeta^n$$

converges in the annulus $1/r_1 > |\zeta| > 1/r_2$, and hence converges absolutely in the disk $|\zeta| < 1/r_1$ (where if $r_1 = 0$, the disk $|\zeta| < 1/r_1$ is taken to be $\mathbf{C}$). Thus

$$\sum_{n=1}^{\infty} a_{-n}(\zeta - z_0)^{-n}$$

converges absolutely on the annulus $r_1 < |\zeta - z_0| < r_2$ and the theorem is proved.

Corollary 5.1 Let $0 \le r_1 < r_2 \le \infty$, let $z_0 \in \mathbf{C}$, and let f be analytic on the annulus $r_1 < |z - z_0| < r_2$. Then there exist unique analytic functions

$$f_1 \colon \{z \colon |z| < r_2\} \to \mathbf{C} \quad \text{and} \quad f_2 \colon \left\{z \colon |z| < \frac{1}{r_1}\right\} \to \mathbf{C}$$

with $f_2(0) = 0$ such that if $r_1 < |z - z_0| < r_2$, then

$$f(z) = f_1(z - z_0) + f_2\left(\frac{1}{z - z_0}\right) \qquad (5.3)$$

In fact,

$$f_1(z) = \sum_{n=0}^{\infty} a_n z^n \qquad |z| < r_2$$

and

$$f_2(z) = \sum_{n=1}^{\infty} a_{-n} z^n \qquad |z| < \frac{1}{r_1}$$

Hence the coefficient expansion for Eq. (5.2) is unique.

Proof. Clearly we need only prove uniqueness, and it is here that the requirement $f_2(0) = 0$ is used. So suppose $\hat{f}_1$ and $\hat{f}_2$ form another such decomposition. Then if $r_1 < |z - z_0| < r_2$, we have

$$[f_1(z - z_0) - \hat{f}_1(z - z_0)] + \left[f_2\left(\frac{1}{z - z_0}\right) - \hat{f}_2\left(\frac{1}{z - z_0}\right)\right] = 0$$

Let $g \colon \mathbf{C} \to \mathbf{C}$ be defined by

$$g(z) = \begin{cases} f_1(z - z_0) - \hat{f}_1(z - z_0) & |z - z_0| < r_2 \\ -\left[f_2\left(\dfrac{1}{z - z_0}\right) - \hat{f}_2\left(\dfrac{1}{z - z_0}\right)\right] & |z - z_0| \ge r_2 \end{cases}$$

Then g is clearly an entire function, and moreover,

$$\lim_{|z| \to \infty} g(z) = -[f_2(0) - \hat{f}_2(0)] = 0$$

Hence g is a bounded entire function and therefore constant. So again, as

$$\lim_{|z| \to \infty} g(z) = 0$$

we see that $g(z) = 0$ for all $z \in \mathbf{C}$, and the corollary is proved.

The functions $\phi_1(z) = f_1(z - z_0)$ and $\phi_2(z) = f_2(1/(z - z_0))$ in Eq. (5.3) have traditionally been called, respectively, the *regular part* of f and the *singular* (or *principal*) *part* of f on the annulus $r_1 < |z - z_0| < r_2$.

Remark 5.1 Note that if $r_1 = 0$ in Corollary 5.1, then the corollary still holds, and f_2 is in fact an entire function.

The following corollary is due to Riemann.

Corollary 5.2 Let $z_0 \in \mathbf{C}$, $r > 0$, and $f: \{z: 0 < |z - z_0| < r\} \to \mathbf{C}$ be a bounded analytic function. Then f has an analytic extension

$$F: \{z: |z - z_0| < r\} \to \mathbf{C}$$

Proof. By Corollary 5.1, there exist analytic functions $f_1: \{z: |z| < r\} \to \mathbf{C}$ and $f_2: \mathbf{C} \to \mathbf{C}$ with $f_2(0) = 0$ such that if $0 < |z - z_0| < r$, then

$$f(z) = f_1(z - z_0) + f_2\left(\frac{1}{z - z_0}\right)$$

By the above equation, f_2 is bounded on the set $\{z: |z| \geq 2/r\}$. Thus since f_2 is an entire function, it is clearly a bounded entire function. Therefore, because $f_2(0) = 0$, we see that $f_2(z) = 0$ for all $z \in \mathbf{C}$, and hence that $f(z) = f_1(z - z_0)$ whenever $0 < |z - z_0| < r$. The desired extension $F: \{z: |z - z_0| < r\} \to \mathbf{C}$ is given by the formula

$$F(z) = f_1(z - z_0)$$

whenever $|z - z_0| < r$.

The fact that the Laurent-series expansion of a function f about a point is unique in a neighborhood of this point enables us in many cases to compute these series in a very easy way without using the integral formula (5.1) for the coefficients.

Example 5.1 Find the Laurent-series expansion about $z_0 = 0$, which represents the following functions in the indicated regions.

a. $\dfrac{10}{(z + 2)(z^2 + 1)}$ for $1 < |z| < 2$

b. $\dfrac{z^2 - 1}{(z + 2)(z + 3)}$ for $|z| > 3$

c. $\dfrac{24}{z^2(z - 1)(z + 2)}$ for $0 < |z| < 1$

Solution

a. By decomposing the function into partial fractions, we obtain

$$\frac{10}{(z + 2)(z^2 + 1)} = \frac{A}{z + 2} + \frac{B}{z + i} + \frac{C}{z - i}$$

where $A = 2$, $B = -1 + 2i$, and $C = -(1 + 2i)$. Now observe that

$$\frac{1}{z + 2} = \frac{1}{2}\frac{1}{(1 + z/2)} = \frac{1}{2}\sum_{n=0}^{\infty}(-1)^n\frac{z^n}{2^n} \quad \text{since } |z| < 2$$

$$\frac{1}{z + i} = \frac{1}{z}\frac{1}{(1 + i/z)} = \sum_{n=0}^{\infty}(-1)^n\frac{i^n}{z^{n+1}} \quad \text{since } |i| < |z|$$

$$\frac{1}{z - i} = \frac{1}{z}\frac{1}{(1 - i/z)} = \sum_{n=0}^{\infty}\frac{i^n}{z^{n+1}} \quad \text{since } |i| < |z|$$

Thus

$$\frac{10}{(z + 2)(z^2 + 1)} = \sum_{n=0}^{\infty} \frac{(-1)^n}{2^n} z^n + (-1 + 2i) \sum_{n=0}^{\infty} (-1)^n i^n z^{-n-1}$$

$$- (1 + 2i) \sum_{n=0}^{\infty} i^n z^{-n-1} \qquad \text{for } 1 < |z| < 2$$

b. Here we have

$$\frac{z^2 - 1}{(z + 2)(z + 3)} = 1 - \frac{5z + 7}{(z + 2)(z + 3)} = 1 + \frac{A}{z + 2} + \frac{B}{z + 3}$$

where $A = 3$ and $B = -8$. Since $|z| > 3$, we have

$$\frac{1}{z + 2} = \frac{1}{z} \frac{1}{(1 + 2/z)} = \sum_{n=0}^{\infty} (-1)^n \frac{2^n}{z^{n+1}}$$

$$\frac{1}{z + 3} = \frac{1}{z} \frac{1}{(1 + 3/z)} = \sum_{n=0}^{\infty} (-1)^n \frac{3^n}{z^{n+1}}$$

Thus

$$\frac{z^2 - 1}{(z + 2)(z + 3)} = 1 + 3 \sum_{n=0}^{\infty} (-1)^n 2^n z^{-n-1} - 8 \sum_{n=0}^{\infty} (-1)^n 3^n z^{-n-1}$$

where $|z| > 3$.

c. We have

$$\frac{24}{z^2(z - 1)(z + 2)} = \frac{A}{z} + \frac{B}{z^2} + \frac{C}{z + 1} + \frac{D}{z + 2}$$

where $A = -6$, $B = -12$, $C = 8$, and $D = -2$. We have

$$\frac{1}{z - 1} = -\frac{1}{1 - z} = -\sum_{n=0}^{\infty} z^n \qquad \text{since } |z| < 1$$

$$\frac{1}{z + 2} = \frac{1}{2} \frac{1}{(1 + z/2)} = \frac{1}{2} \sum_{n=0}^{\infty} (-1)^n \frac{z^n}{2^n} \qquad \text{since } |z| < 2$$

Thus

$$\frac{24}{z^2(z - 1)(z + 2)} = -\frac{6}{z} - \frac{12}{z^2} - 8 \sum_{n=0}^{\infty} z^n - \sum_{n=0}^{\infty} \frac{(-1)^n}{2^n} z^n$$

for $0 < |z| < 1$.

EXERCISES

5.2.1 Find the Laurent-series expansion about $z_0 = 1$ which represents the following functions in the indicated regions.

a. $\dfrac{1}{(z + 1)(z^2 - 2z + 2)}$ in $1 < |z - 1| < 2$

b. $\dfrac{1}{(z + 1)(z + 2)}$ in $|z - 1| > 3$

c. $\dfrac{1}{(z - 1)^2(z - 2)(z + 1)}$ in $0 < |z - 1| < 1$

5.2.2 Find the Laurent-series expansion about $z_0 = 0$ which represents the following functions in the indicated regions.

a. $\dfrac{\sinh z}{z^8}$ in $0 < |z|$

b. $\dfrac{e^{z^2} - 1}{z^3}$ in $0 < |z| < 1$

c. $\dfrac{1}{(z + 2)(z - 3)}$ in $1 < |z| < 2$

d. $\dfrac{2z - 3}{(z - 1)^2(z + i)}$ in $|z| > 1$

5.2.3 Let $z_0 \in \mathbf{C}$, $r > 0$, and $f: \{z: |z - z_0| > r\} \to \mathbf{C}$ be a bounded analytic function. Show that there exists an analytic function

$$h: \{z: |z| < 1/r\} \to \mathbf{C}$$

such that

$$f(z) = h\left(\frac{1}{z - z_0}\right)$$

whenever $|z - z_0| > r$.

5.3
CLASSIFICATION OF ISOLATED
SINGULARITIES

Let f be an analytic function. If $z_0 \notin \mathrm{dmn}\, f$ but there does exist $\varepsilon > 0$ such that f restricted to the punctured disk $0 < |z - z_0| < \varepsilon$ is analytic, then we call z_0 an *isolated singularity* of f.

Let $z_0 \in \mathbf{C}$ such that either f is analytic at z_0 or else z_0 is an isolated singularity of f. Then by Theorem 5.1 (Laurent's theorem), there exists $R > 0$ and complex numbers $a_0, a_{\pm 1}, a_{\pm 2}, \ldots$ such that if $0 < |z - z_0| < R$, then

$$f(z) = \sum_{n=0}^{\infty} a_n(z - z_0)^n + \sum_{n=1}^{\infty} a_{-n}(z - z_0)^{-n}$$

Recall also from Corollary 5.1 that the coefficients $a_0, a_{\pm 1}, a_{\pm 2}, \ldots$ are unique and that there actually exist analytic functions $f_1: \{z: |z| < R\} \to \mathbf{C}$ and $f_2: \mathbf{C} \to \mathbf{C}$ defined by

$$f_1(z) = \sum_{n=0}^{\infty} a_n z^n \qquad |z| < R$$

and

$$f_2(z) = \sum_{n=1}^{\infty} a_{-n} z^n \qquad z \in \mathbf{C}$$

so that

$$f(z) = f_1(z - z_0) + f_2\left(\frac{1}{z - z_0}\right)$$

if $0 < |z - z_0| < R$. The function

$$P_{z_0}(z) = f_2\left(\frac{1}{z - z_0}\right) = \sum_{n=1}^{\infty} a_{-n}(z - z_0)^{-n} \qquad z \in \mathbf{C} - \{z_0\}$$

is called the principal part of f at z_0 and is clearly analytic on $\mathbf{C} - \{z_0\}$.

We are now going to investigate the behavior of f near the point z_0. There are three possibilities.

1. $a_{-n} \neq 0$ for infinitely many $n \in \mathbf{N}$. When this happens, we say z_0 is an *isolated essential singularity* of f.
2. $a_{-n} = 0$ for $n = 1, 2, \ldots$. If f is defined at z_0, then $f(z_0) = a_0$, whereas if f is not defined at z_0, then z_0 is called an *isolated removable singularity* as f can be "made" analytic at z_0 by defining $f(z_0) = a_0$. If m is the least integer n such that $a_n \neq 0$, we say that the *order of f at z_0* is m, and we write $\mathcal{O}(f,z_0) = m$. Note in particular that if $z_0 \in \text{dmn } f$ and $\mathcal{O}(f,z_0) = m > 0$, then $f(z_0) = 0$; in this case we say that *f has a zero of order m* (or of *multiplicity m*) at z_0.
3. There exists a positive integer m such that $a_{-m} \neq 0$, whereas $a_{-n} = 0$ for all $n \in \mathbf{N}$ with $n > m$. Then z_0 is called a *pole of f of order m* (or of *multiplicity m*) and we set $\mathcal{O}(f,z_0) = -m$. We say that z_0 is a *simple pole* if $m = 1$, that z_0 is a *double pole* if $m = 2$, and so on. Note that m is unique and that there exists $R > 0$ and unique complex numbers $a_{-m} \neq 0$, $a_{-m+1}, \ldots$ such that

$$f(z) = \sum_{n=-m}^{\infty} a_n(z - z_0)^n \qquad \text{whenever } 0 < |z - z_0| < R \qquad (5.4)$$

We say that f is *meromorphic* at z_0 if f is analytic at z_0 or has a removable singularity at z_0 or has a pole at z_0. If the domain of f is contained in A, then we say that f is *meromorphic on A* if $z_0 \in A$ implies that f is meromorphic at z_0. In this chapter, we shall be primarily interested in meromorphic functions.

Theorem 5.2 Let $R > 0$, let $z_0 \in \mathbf{C}$, let m be a positive integer, and let f be analytic on the punctured disk $0 < |z - z_0| < R$. Then z_0 is a pole of f of order m if and only if $\lim_{z \to z_0} (z - z_0)^m f(z)$ exists and is different from zero.

Proof. If z_0 is a pole of order m, then by examining Eq. (5.4) we see that $\lim_{z \to z_0} (z - z_0)^m f(z) = a_{-m} \neq 0$. So suppose that $\lim_{z \to z_0} (z - z_0)^m f(z)$ exists and is different from zero. Recall that

$$f(z) = f_1(z - z_0) + f_2\left(\frac{1}{z - z_0}\right)$$

where $f_1\colon \{z\colon |z| < R\} \to \mathbf{C}$ and $f_2\colon \mathbf{C} \to \mathbf{C}$ are analytic. So for some $\lambda \in \mathbf{C}$ with $\lambda \neq 0$, we have

$$\lambda = \lim_{z \to z_0} (z - z_0)^m f(z)$$

$$= \lim_{z \to z_0} (z - z_0)^m f_2\left(\frac{1}{z - z_0}\right)$$

$$= \lim_{|z| \to \infty} \frac{f_2(z)}{z^m}$$

So if $n > m$, clearly

$$\lim_{|z| \to \infty} \frac{f_2(z)}{z^n} = 0$$

Suppose $n > m$. We shall show that $a_{-n} = 0$. As

$$f_2(z) = \sum_{n=1}^{\infty} a_{-n} z^n \qquad \text{for } z \in \mathbf{C}$$

we have by Corollary 4.3 that

$$|a_{-n}| = \left| \frac{1}{2\pi i} \int_{|z|=r} \frac{f_2(z)}{z^{n+1}} \, dz \right|$$

$$\leq \frac{2\pi r}{2\pi} \max_{|z|=r} \left| \frac{f_2(z)}{z^{n+1}} \right|$$

$$= \max_{|z|=r} \left| \frac{f_2(z)}{z^n} \right| \to 0 \qquad \text{as } r \to \infty$$

Hence $a_{-n} = 0$ if $n > m$. Using this together with Eq. (5.4), we see that

$$0 \neq \lambda = \lim_{z \to z_0} (z - z_0) f(z) = a_{-m}$$

which proves the theorem.

Corollary 5.3 Let f be a meromorphic function on a connected set A. Suppose f is not identically zero on A. Then the following properties hold.

1. Let $z_0 \in A$ and let $m \in \mathbf{Z}$. Then $\mathcal{O}(f, z_0) = m$ if and only if

$$\lim_{z \to z_0} (z - z_0)^{-m} f(z)$$

   exists and is not equal to zero.
2. Let $z_0 \in A$ and let $m \in \mathbf{Z}$. Then f is meromorphic at z_0 with $\mathcal{O}(f, z_0) = m$ if and only if there exist $R > 0$ and ϕ analytic at z_0 with $\phi(z_0) \neq 0$ such that if $0 < |z - z_0| < R$, then $f(z) = (z - z_0)^m \phi(z)$.
3. A is open.
4. $\{z\colon f(z) = 0\}$ is a discrete subset of A with no limit points in A.
5. $\{z\colon z \text{ is a pole of } f\}$ is a discrete subset of A with no limit points in A.
6. $1/f$ is meromorphic on A, and in fact $\mathcal{O}(1/f, z_0) = -\mathcal{O}(f, z_0)$ for all $z_0 \in A$.

Note in the above corollary that if $\mathcal{O}(f,z_0) = m > 0$, then z_0 is a zero of f of multiplicity m, while if $\mathcal{O}(f,z_0) = m < 0$, then z_0 is a pole of f of order $-m$.

The following remark is often quite useful in classifying the isolated singularities of an analytic function. Part (a) is merely a reformulation of Corollary 5.2 (Riemann's theorem), while part (b) is an easy consequence of Theorem 5.2.

Remark 5.2 Let z_0 be an isolated singularity of f. Then (a) f has a removable singularity at z_0 if and only if f is bounded in a neighborhood of z_0; (b) f has a pole at z_0 if and only if $\lim_{|z| \to \infty} |f(z)| = \infty$.

Example 5.2 Locate and classify all singularities of the following functions.

a. $f(z) = \dfrac{2}{(z-3)^2} + \dfrac{1}{z-3} + e^z$

b. $f(z) = \sin z + \sin \dfrac{1}{z}$

c. $f(z) = \dfrac{\cos z}{z - \pi/2}$

d. $f(z) = \dfrac{\sin z}{z^3}$

Solution

a. It is clear that 3 is the only singularity of f. It is an isolated singularity which is a pole of order 2.
b. As $f(z) = (z - z^3/3! + \cdots) + (1/z - 1/(3!\,z^3) + \cdots)$, we see that 0 is the only singularity of f and is an isolated essential singularity.
c. The only singularity of f is the removable singularity $\pi/2$. If we define $f(\pi/2) = -1$ $[f(\pi/2) = \cos'(\pi/2) = -\sin(\pi/2) = -1]$, then f becomes an entire function.
d. We have

$$f(z) = \frac{1}{z^3}\left(z - \frac{z^3}{3!} + \frac{z^5}{5!} - \cdots\right) = \frac{1}{z^2} - \frac{1}{3!} + \frac{z^2}{5!} - \cdots$$

Hence 0 is the only singularity of f and is a pole of order 2.

Example 5.3 The function

$$f(z) = \frac{z^3 - 8}{(z-2)^2(z+i)^4}$$

has a simple pole at 2 and a pole of order 4 at $-i$.

Proof. In fact,

$$\lim_{z \to 2} (z - 2) \frac{z^3 - 8}{(z - 2)^2(z + i)^4} = \lim_{z \to 2} \frac{z^2 + 2z + 4}{(z + i)^4} = \frac{12}{(2 + i)^4} \neq 0$$

while

$$\lim_{z \to -i} (z + i)^4 \frac{z^3 - 8}{(z - 2)^2(z + i)^4} = \lim_{z \to -i} \frac{z^3 - 8}{(z - 2)^2} = \frac{i - 8}{-i - 2} \neq 0$$

Theorem 5.3 Let f be meromorphic on A, and let K be a compact subset of A. Let $\mathscr{P} = \{w : w \in K \text{ and } w \text{ is a pole of } f\}$. Then $\mathscr{P}$ is a finite set. Moreover, there exists an analytic function $g : \text{dmn } f \cup \mathscr{P} \to \mathbf{C}$ which is meromorphic on A such that if $z \in \text{dmn } f$, then

$$f(z) = g(z) + \sum_{w \in \mathscr{P}} P_w(z)$$

Proof. For each $z \in K$, there exists $R_z > 0$ and complex numbers $a_{\mathcal{O}(f,z)}$, $a_{\mathcal{O}(f,z)+1}, \ldots$ such that if $0 < |w - z| < R_z$, then

$$f(w) = \sum_{n = \mathcal{O}(f,z)}^{\infty} a_n(w - z)^n$$

Hence we see that in $N_{R_z}(z)$, z is the only possible pole of f. As K is compact, there exist $z_1, \ldots, z_n \in K$ such that $K \subset \bigcup_{k=1}^{n} N_{R_{z_k}}(z_k)$, and hence $\mathscr{P} \subset \{z_1, \ldots, z_n\}$. So by Remark 5.2a, $w_0 \in \mathscr{P}$ implies that w_0 is a removable singularity of $f - \sum_{w \in \mathscr{P}} P_w$, which proves the theorem.

The following theorem states that polynomials are those entire functions having a "pole at infinity."

Theorem 5.4 Let $f : \mathbf{C} \to \mathbf{C}$ be an entire function. Then f is a nonconstant polynomial if and only if $\lim_{|z| \to \infty} |f(z)| = \infty$.

Proof. Suppose f is a nonconstant polynomial. By Corollary 4.4, there exist $\lambda, \mu_1, \ldots, \mu_n \in \mathbf{C}$ such that

$$f(z) = \lambda(z - \mu_1) \cdots (z - \mu_n) \qquad \text{for all } z \in \mathbf{C}$$

and hence

$$|f(z)| = |\lambda| \, |z - \mu_1| \cdots |z - \mu_n| \geq |\lambda|(|z| - |\mu_1|) \cdots (|z| - |\mu_n|) \to \infty$$

as $|z| \to \infty$.

Now suppose $|f(z)| \to \infty$ as $|z| \to \infty$. We shall show that f is a nonconstant polynomial. Since f is an entire function, there exist complex numbers $a_0, a_1, \ldots$ such that $f(z) = \sum_{n=0}^{\infty} a_n z^n$ for all $z \in \mathbf{C}$. Since $|f(z)| \to \infty$ as $|z| \to \infty$, it follows by Remark 5.2 that if we let $g(z) = f(1/z)$ whenever $z \in \mathbf{C} - \{0\}$, then g has a pole at zero. So as

$$g(z) = f\left(\frac{1}{z}\right) = \sum_{n=0}^{\infty} a_n \frac{1}{z^n}$$

is the Laurent-series expansion for g about 0, we see there exists a positive integer $m = -\mathcal{O}(g,0)$ such that if $n \geq m$, then $a_n = 0$. Hence $f(z) = \sum_{n=0}^{m} a_n z^n$ for $z \in \mathbf{C}$, which proves the theorem.

Remark 5.2 described the nature of an analytic function in a neighborhood of an isolated, nonessential singularity. The following theorem, called the Casorati-Weierstrass theorem (Felice Casorati, 1835–1890, Italian), describes the behavior of an analytic function in a neighborhood of an isolated singularity.

Theorem 5.5. *Casorati-Weierstrass Theorem* Let z_0 be an isolated singularity of f. Then z_0 is an isolated essential singularity of f if and only if $w \in \mathbf{C}$ implies that there exists a sequence of complex numbers $z_1, z_2, \ldots$ such that $\lim_{n \to \infty} z_n = z_0$ and $\lim_{n \to \infty} f(z_n) = w$. That is, the image of f near z_0 is dense in $\mathbf{C}$.

Proof. If $w \in \mathbf{C}$ implies the existence of a sequence of complex numbers $z_1, z_2, \ldots$ satisfying $\lim_{n \to \infty} z_n = z_0$ and $\lim_{n \to \infty} f(z_n) = w$, then by Remark 5.2, we know that z_0 can be neither a removable singularity nor a pole and hence must be an essential singularity. So suppose z_0 is an essential singularity of f and let $w \in \mathbf{C}$. It suffices to show that there exists a sequence of complex numbers $z_1, z_2, \ldots$ such that $\lim_{n \to \infty} z_n = z_0$ and $\lim_{n \to \infty} f(z_n) = w$. So suppose there is no such sequence. We shall derive a contradiction. Since for every sequence $z_1, z_2, \ldots$ in $\mathbf{C}$ with $\lim_{n \to \infty} z_n = z_0$ the existence of $\lim_{n \to \infty} f(z_n)$ implies $\lim_{n \to \infty} f(z_n) \neq w$, we see that there exists $\varepsilon > 0$ and $\delta > 0$ such that $0 < |z - z_0| < \delta$ implies $|f(z) - w| > \varepsilon$. Hence we see that if we define $g: \{z: 0 < |z - z_0| < \delta\} \to \mathbf{C}$ by

$$g(z) = \frac{1}{f(z) - w} \qquad \text{if } 0 < |z - z_0| < \delta$$

then g has a removable singularity at z_0. So by Corollary 5.3, we know that $1/g$ (and hence f) is meromorphic at z_0. Thus z_0 cannot be an essential singularity of f. This contradiction proves the theorem.

Until this point in the section, we have been interested only in isolated singularities. We shall conclude the section with a few remarks about singularities in general.

FELICE CASORATI

Felice Casorati, one of the first Italian analysts, was born in Pavia, Italy, on December 17, 1835. (Geronimo Cardano was also born in Pavia, in 1501.) Casorati died in Casteggio, Italy, on September 11, 1890.

In Sec. 4.5, we defined a nonremovable singularity of a power series to be a point on the circle of convergence about which it is impossible to analytically extend the series. With this in mind, we make the following definition. Let f be an analytic function and let $z_0 \in$ bd $(\text{dmn } f)$. Then if there exists an open set O with $z_0 \in O$ and an analytic extension F: dmn $f \cup O \to \mathbf{C}$ of f (in other words, an analytic function F: dmn $f \cup O \to \mathbf{C}$ such that $F | \text{dmn } f = f$), we call z_0 a *removable singularity* of f. If $z_0 \in$ bd $(\text{dmn } f)$ is not a removable singularity of f, we call z_0 a *nonremovable singularity* of f. Note that if z_0 is a singularity of f, then while $z_0 \notin \text{dmn } f$ [since f is analytic, dmn f is open, and thus dmn $f \cap$ bd $(\text{dmn } f) = \varnothing$], z_0 does have the property that every ε neighborhood of it contains points at which f is analytic. Recall that a singularity z_0 of f was called an isolated singularity of f if there existed $\varepsilon > 0$ so that f when restricted to the punctured disk $0 < |z - z_0| < \varepsilon$ was analytic. A nonremovable singularity z_0 of f which is not isolated is called an *essential* singularity (this use of the word essential is in addition to our use at the beginning of this section).

Example 5.4 Study the singularities of the following functions.

a. $f(z) = \dfrac{1}{\cos 1/z}$

b. $f(z) = z \qquad$ if $|z| < 1$

c. $f(z) = \displaystyle\sum_{n=1}^{\infty} nz^n$

Solution

a. The points

$$\left\{ \frac{1}{n\pi + \pi/2} : n \in \mathbf{Z} \right\}$$

are isolated singularities of f, whereas 0 is a singularity of f that is not isolated, since $1/(n\pi + \pi/2) \to 0$ as $n \to \infty$. Hence 0 is an essential singularity of f. By Theorem 5.2 we see that $1/(n\pi + \pi/2)$ is a simple pole for all $n \in \mathbf{Z}$.

b. The points $\{z: |z| = 1\}$ are all removable, nonisolated singularities.

c. If $|z| < 1$, let $g(z) = \sum_{n=1}^{\infty} z^n$. Then if $|z| < 1$,

$$\frac{1}{1 - z} = \sum_{n=0}^{\infty} z^n = g(z)$$

and hence

$$f(z) = \sum_{n=0}^{\infty} nz^n = g'(z) = \frac{1}{(1 - z)^2}$$

Thus if $|z| = 1$ and $z \neq 1$, then z is a removable singularity, while 1 is a nonremovable singularity.

The reader may be disturbed by our saying that the function in part (b) above has singularities, and might wish to "make the domain of the function f as big as possible before determining the singularities of f." This would work in parts (b) and (c); however, no sense can be made of the phrase "make the domain as big as possible" in the following example, and so we are stuck with our definition.

Example 5.5 Study the singularities of the function

$$f(z) = \sum_{n=1}^{\infty} \frac{(-1)^n}{n} z^n \qquad |z - 1| < 1$$

Solution. $f(z) = \text{Log } z$ if $|z - 1| < 1$. Thus the removable singularities are the points $\{z : |z - 1| = 1, z \neq 0\}$. It is impossible to define log to be single-valued in any neighborhood of 0, and so 0 is a nonremovable singularity of f. It follows by Corollary 4.12 that f has unique analytic extensions

$$f_1 : \mathbf{C} - \{x : x \leq 0\} \to \mathbf{C}$$

and

$$f_2 : \mathbf{C} - \{iy : y \leq 0\}$$

One can check to see that, in the third quadrant $\{x + iy : x < 0 \text{ and } y < 0\}$, f_1 and f_2 do not agree, but they do agree in the disk $|z - 1| < 1$. Thus there is no unique way of extending the domain of f, and so we cannot "make the domain of f as big as possible."

Example 5.6 Study the singularities of the function

$$f(z) = \text{Log } z \qquad \text{if } z \in \mathbf{C} - \{x : x \leq 0\}$$

Solution. Clearly every point of $\{x : x \leq 0\}$ is a nonisolated essential singularity, since Log cannot even be made continuous at any point of $\{x : x \leq 0\}$. In Example 4.8, we showed that if $|z - (-2 + i)| < 1$, then

$$\text{Log } z = \frac{1}{2} \text{Log } 5 + i \text{ Arg } (-2 + i) - \sum_{n=1}^{\infty} \frac{1}{n} \left(\frac{2 + i}{5} \right)^n [z - (-2 + i)]^n$$

$$(5.5)$$

where the right-hand side had radius of convergence $\sqrt{5} = |(-2 + i) - 0|$. Thus while the radius of convergence of the power series is the distance from the center $-2 + i$ to the nearest nonremovable singularity 0 of the power series, it is not the distance from $-2 + i$ to the nearest nonremovable singularity of the function being represented which is Log; the nonremovable singularity of Log nearest to $-2 + i$ is -2, and the distance between $-2 + i$ and -2 is 1.

If f is analytic at z_0 and has nonremovable singularities, then there will exist a nonremovable singularity w of f such that if λ is any other nonremovable singularity of f, then $|z_0 - w| \leq |z_0 - \lambda|$. Thus it does make sense to speak of the nearest nonremovable singularity. If R is the radius of convergence of the power series representing f at z_0, then all we can say in general is that $|z_0 - w| \leq R$, as the above example illustrated.

EXERCISES

5.3.1 Locate and classify all singularities of the following functions. If the singularity is a pole, find its order.

a. $f(z) = \dfrac{1}{\sin z}$

b. $f(z) = \dfrac{1 - \cos z}{z^6}$

c. $f(z) = \dfrac{1}{-1 + e^z}$

d. $f(z) = e^{1/z}$

5.3.2 The proof of the Casorati-Weierstrass theorem used only the fact that if z_0 is a pole of f of order m, then $\lim_{z \to z_0} (z - z_0)^m f(z)$ exists and is not zero, but it did not use the converse. Use the Casorati-Weierstrass theorem to prove the converse.

5.3.3 Show that the function $f(z) = (1 - \cos z)/z^2$ has a removable singularity at $z = 0$. How does one extend f to an entire function?

5.3.4 Let f be analytic on D. Let R be the set of isolated removable singularities of f. Show by means of an argument similar to that in the proof of Theorem 5.3 that if K is any compact set, then $R \cap K$ is finite and hence R is at most countable. Use this fact to prove that there exists a unique analytic extension $F: D \cup R \to \mathbf{C}$ of f having no isolated removable singularities.

5.4

THE CAUCHY RESIDUE THEOREM

Let f be meromorphic on A. Then if $z_0 \in A$, there exists $R > 0$ and unique complex numbers $a_{\mathcal{O}(f,z_0)}, a_{\mathcal{O}(f,z_0)+1}, \ldots$ such that if $0 < |z - z_0| < R$, then

$$f(z) = \sum_{n=\mathcal{O}(f,z_0)}^{\infty} a_n(z - z_0)^n$$

If $\mathcal{O}(f,z_0) \geq 0$, we define the *residue* of f at z_0 to be

$$\text{Res }(f,z_0) = 0$$

but if $\mathcal{O}(f,z_0) < 0$, we define the residue of f at z_0 to be

$$\text{Res }(f,z_0) = a_{-1}$$

Recall that in either case, if $0 < r < R$ and $\gamma: [0,1] \to \mathbf{C}$ is the curve defined by $\gamma(t) = z_0 + re^{2\pi it}$ if $0 \le t \le 1$, then

$$\text{Res }(f,z_0) = \frac{1}{2\pi i} \int_\gamma f(\zeta)\, d\zeta$$

The following theorem is often quite useful in computing the residue of a function f at a pole z_0 of f.

Theorem 5.6 Let z_0 be a pole of f of order m. Then

$$\text{Res }(f,z_0) = \frac{1}{(m-1)!} \lim_{z \to z_0} \frac{d^{m-1}}{dz^{m-1}} [(z-z_0)^m f(z)] \qquad (5.6)$$

where the notation $d^{m-1}[g(z)]/dz^m$ means $g^{(m-1)}(z)$, the $(m-1)$st derivative of g at z.

Proof. Let R, $a_{\mathcal{O}(f,z_0)}$, $a_{\mathcal{O}(f,z_0)+1}, \ldots$ be as above. As z_0 is a pole of f of order m, we see that $\mathcal{O}(f,z_0) = -m < 0$. So if $0 < |z - z_0| < R$, we have

$$(z-z_0)^m f(z) = (z-z_0)^m \sum_{n=-m}^{\infty} a_n(z-z_0)^n = \sum_{n=0}^{\infty} a_{n-m}(z-z_0)^n$$

Therefore,

$$\text{Res }(f,z_0) = a_{-1} = \frac{1}{(m-1)!} \lim_{z \to z_0} \frac{d^{m-1}}{dz^{m-1}} [(z-z_0)^m f(z)]$$

Example 5.7 Compute the residues of the function

$$f(z) = \frac{\sin z}{(z-i)(z+2)^2}$$

at its singularities.

Solution. Clearly i and -2 are f's only singularities. It is also clear that i is a pole of order 1 and that -2 is a pole of order 2. Thus

$$\text{Res }(f,i) = \lim_{z \to i} (z-i) \frac{\sin z}{(z-i)(z+2)^2} = \frac{\sin i}{(i+2)^2}$$

and

$$\text{Res }(f,-2) = \lim_{z \to -2} \frac{d}{dz} \left[(z+2)^2 \frac{\sin z}{(z-i)(z+2)^2} \right]$$

$$= \lim_{z \to -2} \frac{(z-i)\cos z - \sin z}{(z-i)^2}$$

$$= \frac{(-2-i)\cos 2 + \sin 2}{3+4i}$$

The heart of this chapter is the Cauchy residue theorem. In order to give a simple formulation of its statement (which is still general enough to include the great bulk of applications), we are first going to state without proof two "geometrically obvious" (but not so easy to prove) results dealing with simple closed curves in the plane. The first theorem is called the Jordan curve theorem (Camille Jordan, 1838–1922, French).

Theorem 5.7 Let γ be a simple closed curve in $\mathbf{C}$. Then γ divides $\mathbf{C}$ into two disjoint regions, one of which is bounded and simply connected and called the *interior* of γ, the other unbounded and called the *exterior* of γ.

Because of the importance of this theorem, it is customary to call a simple closed curve a *Jordan curve*.

Theorem 5.8 Let γ_1 and γ_2 be two Jordan curves in a simply connected domain D, and suppose that $z_0 \in$ interior $\gamma_1 \subset$ interior γ_2. Then γ_1 is homotopic through closed curves (actually through Jordan curves) in $D - \{z_0\}$ either to γ_2 or else to $\tilde{\gamma}_2$.

We are now ready to discuss the concept of the *oriented Jordan curve*. Let γ be a Jordan curve and let $z_0 \in$ interior γ. Then since interior γ is an open set, there exists $r > 0$ such that the closed disk $|z - z_0| \le r$ is contained in interior γ. Thus if we let $\Gamma(t) = z_0 + re^{it}$ for $0 \le t \le 2\pi$, then Theorem 5.8 implies either $I(\gamma, z_0) = I(\Gamma, z_0) = 1$ or else $I(\gamma, z_0) = I(\tilde{\Gamma}, z_0) = -1$. So as interior γ is connected, we see that either $z \in$ interior γ implies $I(\gamma, z) = 1$, or else $z \in$ interior γ implies $I(\gamma, z) = -1$. We call γ a *positively oriented* Jordan curve if $I(\gamma, z) = 1$ for all $z \in$ interior γ, while we call γ a *negatively oriented* Jordan curve if $I(\gamma, z) = -1$ for all $z \in$ interior γ. So we see that if γ is a simple closed curve, then γ is either a positively oriented simple closed curve or a negatively oriented simple closed curve, and moreover that if γ is a negatively oriented Jordan curve, then $\tilde{\gamma}$ is a positively oriented Jordan curve.

Theorem 5.9. *The Cauchy Residue Theorem* Let D be a simply connected domain, and let γ be a positively oriented simple closed curve in D. Let $z_1, \ldots, z_n$ be points in the interior of γ, and let $f: D - \{z_1, \ldots, z_n\} \to \mathbf{C}$ be analytic with poles at $z_1, \ldots, z_n$. Then

$$\int_\gamma f(z)\, dz = 2\pi i \sum_{k=1}^{n} \text{Res}\,(f, z_k)$$

Proof. If for $k = 1, \ldots, n$ we have P_{z_k} as the principal part of f at z_k, then clearly the function g defined by

$$g(z) = f(z) - \sum_{k=1}^{n} P_{z_k}\left(\frac{1}{z - z_k}\right)$$

Fig. 5.2

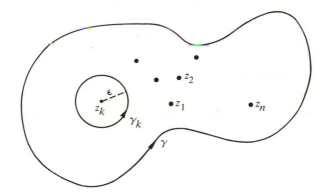

is analytic on D. Thus by Corollary 4.10, we have

$$0 = \int_\gamma g(z)\, dz = \int_\gamma f(z)\, dz - \sum_{k=1}^n \int_\gamma P_{z_k}\left(\frac{1}{z - z_k}\right) dz \qquad (5.7)$$

Let k be an integer with $1 \le k \le n$. It suffices to show that

$$\int_\gamma P_{z_k}\left(\frac{1}{z - z_k}\right) dz = 2\pi i \operatorname{Res}(f, z_k)$$

Let $\varepsilon > 0$ be sufficiently small so that the positively oriented circle γ_k centered at z_k is contained in the interior of γ and contains none of the points z_j for $j \ne k$. See Fig. 5.2.

Then because γ is positively oriented, it follows by Theorem 5.8 that γ is homotopic through closed curves in $D - \{z_k\}$ to γ_k and hence that

$$\int_\gamma P_{z_k}\left(\frac{1}{z - z_k}\right) dz = \int_{\gamma_k} P_{z_k}\left(\frac{1}{z - z_k}\right) dz = 2\pi i \operatorname{Res}(f, z_k)$$

The proof is complete.

Example 5.8 Prove that

$$\int_\gamma \frac{\cos z}{z(z - 1)^2}\, dz = 2\pi i (1 - \sin 1 - \cos 1) \qquad \text{where } \gamma\colon |z| = 3$$

Proof. The only singularities of the integrand

$$f(z) = \frac{\cos z}{z(z - 1)^2}$$

inside γ are $z = 0$, which is a simple pole, and $z = 1$, which is a double pole. Thus by Theorem 5.9, we have

$$\int_\gamma \frac{\cos z}{z(z - 1)^2}\, dz = 2\pi i [\operatorname{Res}(f, 0) + \operatorname{Res}(f, 1)]$$

By Theorem 5.6 we have

$$\text{Res } (f,0) = \lim_{z \to 0} z \, \frac{\cos z}{z(z-1)^2} = 1$$

and

$$\text{Res } (f,1) = \frac{1}{1!} \lim_{z \to 1} \frac{d}{dz} \left[(z-1)^2 \, \frac{\cos z}{z(z-1)^2} \right]$$

$$= \lim_{z \to 1} \frac{d}{dz} \left(\frac{\cos z}{z} \right) = -\cos 1 - \sin 1$$

Hence

$$\int_\gamma \frac{\cos z}{z(z-1)^2} \, dz = 2\pi i (1 - \cos 1 - \sin 1)$$

Example 5.9 Evaluate the integral

$$I = \int_\gamma \frac{z^3 - 1}{(z+1)(z^2 - 4)(z+i)} \, dz \qquad \text{where } \gamma: |z+1| = \tfrac{3}{2}$$

Solution. The only singularities of the integrand

$$f(z) = \frac{z^3 - 1}{(z+1)(z^2 - 4)(z+i)}$$

are -1, ± 2, and $-i$ of which -1, -2, and $-i$ lie in the interior of γ and are simple poles, while the other singularity $z = 2$ lies in the exterior of γ. Thus $I = 2\pi i [\text{Res } (f,-1) + \text{Res } (f,-2) + \text{Res } (f,-i)]$. By Theorem 5.6 we get

$$\text{Res } (f,-1) = \lim_{z \to -1} (z+1) \frac{z^3 - 1}{(z+1)(z^2 - 4)(z+i)} = \frac{-2}{(-3)(-1+i)}$$

$$= -\frac{1+i}{3}$$

$$\text{Res } (f,-2) = \lim_{z \to -2} (z+2) \frac{z^3 - 1}{(z+1)(z^2 - 4)(z+i)}$$

$$= \frac{-9}{(-1)(-4)(-2+i)} = \frac{9(2+i)}{20}$$

$$\text{Res } (f,-i) = \lim_{z \to -1} (z+i) \frac{z^3 - 1}{(z+1)(z^2 - 4)(z+i)}$$

$$= \frac{-i-1}{(1-i)(-5)} = \frac{1}{5}$$

Hence, by Theorem 5.9 we have

$$I = 2\pi i \left[-\frac{1-i}{3} + \frac{9(2+i)}{20} + \frac{1}{5} \right] = \frac{-7 + 46i}{30} \pi$$

The following result is the generalized version of the Cauchy residue theorem which we are going to prove.

Theorem 5.10. *Generalized Cauchy Residue Theorem* Let f be meromorphic on A, and let $\gamma\colon [a,b] \to$ dmn f be a closed curve which is contractible in A. Then $\mathscr{P} = \{z\colon z \in A,\ I(\gamma,z) \neq 0,\ \text{and Res } (f,z) \neq 0\}$ is a finite set, and

$$\int_\gamma f(z)\, dz = 2\pi i \sum_{z \in \mathscr{P}} I(\gamma,z)\, \text{Res } (f,z)$$

Proof. Let H be a homotopy through closed curves in A between γ and a constant curve. By Corollary 4.14 we have $\{z\colon z \in A \text{ and } I(\gamma,z) \neq 0\} \subset \text{Im } H$, which is a compact set. Hence we have $\mathscr{P} \subset \text{im } H \cap \{z\colon z \text{ is a pole of } f\}$. So it follows by Theorem 5.3 that $\mathscr{P}$ is finite and that there exists an analytic function $g\colon \mathscr{P} \cup \text{dmn } f \to \mathbf{C}$ such that if $z \in \mathscr{P} \cup \text{dmn } f$, then $f(z) = g(z) + \sum_{w \in \mathscr{P}} P_w[1/(z - w)]$. Thus

$$\int_\gamma f(z)\, dz = \int_\gamma g(z)\, dz + \sum_{w \in \mathscr{P}} \int_\gamma P_w\left(\frac{1}{z - w}\right) dz$$

As γ is a closed curve which is null-homotopic in dmn g, we have $\int_\gamma g(z)\, dz = 0$. Since

$$\int_\gamma P_w\left(\frac{1}{z - w}\right) dz = 2\pi i I(\gamma,w)\, \text{Res } (f,w)$$

the theorem follows.

Example 5.10 Let

$$\gamma(t) = \begin{cases} 4 \cos t + 3i \sin t & \text{for } 0 \leq t \leq 4\pi \\ 4 + 3i + 3e^{-i(t - 7\pi/2)} & \text{for } 4\pi \leq t \leq 10\pi \end{cases}$$

and let $f(z) = 1/[(z - 3 - 2i)(z - 4 - 3i)]$. See Fig. 5.3. Compute $\int_\gamma f(z)\, dz$.

Solution. By Theorem 5.10, we have

$$\int_\gamma f(z)\, dz$$

$$= 2\pi i[I(\gamma,3 + 2i)\, \text{Res } (f,3 + 2i) + I(\gamma,4 + 3i)\, \text{Res } (f,4 + 3i)]$$

$$= 2\pi i\left[(-1)\frac{1}{-1 - 2i} + (-3)\frac{1}{1 + 2i}\right] = -\frac{4\pi i}{1 + 2i}$$

EXERCISES

5.4.1 Evaluate the integral

$$\int_\gamma \frac{(1 - z^4)e^{2z}}{z^3}\, dz \qquad \text{where } \gamma\colon |z| = {}^1/_2$$

Fig. 5.3

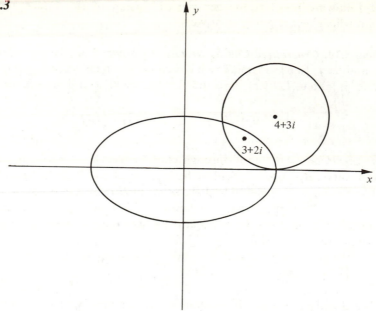

5.4.2 Evaluate the integral

$$\int_\gamma \frac{\sinh z}{\sin z}\, dz \qquad \text{where } \gamma\colon |z| = 1$$

5.4.3 Evaluate the integral

$$\int_\gamma \frac{3z\, dz}{(z^2 + 1)(z^2 + z + 1)^2} \qquad \text{where } \gamma\colon |z - i| = 1$$

5.4.4 Let

$$\gamma(t) = \begin{cases} e^{2\pi i t} & \text{for } 0 \le t \le 1 \\ 2 + e^{\pi i(1 - 2t)} & \text{for } 1 \le t \le 2 \\ 2 + e^{\pi i(1 + 2t)} & \text{for } 2 \le t \le 3 \end{cases}$$

and let

$$f(z) = \frac{1}{(z - \frac{1}{2})[z - (2 + i/2)]^{10}[z - (-3 + i)]^{15}}$$

Find $\int_\gamma f(z)\, dz$.

5.5

**APPLICATIONS OF CAUCHY'S RESIDUE
THEOREM TO THE EVALUATION OF
REAL INTEGRALS**

Our first application will be the development of
a technique often useful in the integration of rational functions.

We call a function $P\colon \mathbf{C} \times \mathbf{C} \to \mathbf{C}$ a *polynomial of two variables of degree*

n if there exist complex numbers $a_{(0,0)}, a_{(0,1)}, a_{(1,0)}, \ldots, a_{(0,k)}, a_{(1,k-1)}, \ldots,$ $a_{(k,0)}, \ldots, a_{(0,n)}, a_{(1,n-1)}, \ldots, a_{(n,0)}$ with

$$\sum_{p=0}^{n} |a_{(p,n-p)}| \neq 0$$

such that

$$P(z,w) = \sum_{k=0}^{n} \sum_{p=0}^{k} a_{(p,k-p)} z^p w^{k-p} \qquad (z,w) \in \mathbf{C} \times \mathbf{C}$$

We call F a *rational function of two variables* if there exist polynomials of two variables P and Q (Q not identically zero) such that

$$F(z,w) = \frac{P(z,w)}{Q(z,w)} \qquad \text{if } (z,w) \in \mathbf{C} \times \mathbf{C} \qquad \text{and} \qquad Q(z,w) \neq 0$$

For example, $(3z^2 + 5zw)/(w^2 - zw)$ and zw are rational functions.

Theorem 5.11 Let F be a rational function such that $F(\cos t, \sin t)$ is defined for all $t \in [0, 2\pi]$. Then

$$\int_0^{2\pi} F(\cos t, \sin t)\, dt = 2\pi i \sum_{k=1}^{n} \mathrm{Res}\,(f, z_k)$$

where

$$f(z) = \frac{1}{iz} F\left(\frac{z^2 + 1}{2z}, \frac{z^2 - 1}{2iz} \right) \qquad \text{and} \qquad z_1, z_2, \ldots, z_n$$

are the poles of f inside the unit circle $|z| = 1$.

Proof. If we let $\gamma(t) = e^{it}$ for $0 \leq t \leq 2\pi$, then

$$\cos t = \frac{e^{it} + e^{-it}}{2} = \frac{1}{2}\left[\gamma(t) + \frac{1}{\gamma(t)} \right]$$

$$\sin t = \frac{e^{it} - e^{-it}}{2i} = \frac{1}{2i}\left[\gamma(t) - \frac{1}{\gamma(t)} \right]$$

and $\gamma'(t) = ie^{it} = i\gamma(t)$. Hence

$$\int_0^{2\pi} F(\cos t, \sin t)\, dt = \int_0^{2\pi} F\left\{ \frac{[\gamma(t)]^2 + 1}{2\gamma(t)}, \frac{[\gamma(t)]^2 - 1}{2i\gamma(t)} \right\} dt$$

$$= \int_0^{2\pi} \frac{1}{i\gamma(t)} F\left\{ \frac{[\gamma(t)]^2 + 1}{2\gamma(t)}, \frac{[\gamma(t)]^2 - 1}{2i\gamma(t)} \right\} \gamma'(t)\, dt$$

$$= \int_0^{2\pi} f(\gamma(t))\gamma'(t)\, dt$$

$$= \int_\gamma f(z)\, dz$$

$$= 2\pi i \sum_{k=1}^{n} \mathrm{Res}\,(f, z_k)$$

Note that in the above, if we write symbolically $z = e^{it}$, then our substitution is

$$\cos t = \frac{e^{it} + e^{-it}}{2} = \frac{1}{2}\left(z + \frac{1}{z}\right) = \frac{z^2 + 1}{2z}$$

$$\sin t = \frac{e^{it} - e^{-it}}{2i} = \frac{z^2 - 1}{2iz}$$

and $dz = ie^{it}\, dt = iz\, dt$; that is, $dt = dz/iz$.

Example 5.11 Show that

$$\int_0^{2\pi} \frac{dt}{\sqrt{5} + \cos t} = \pi$$

Proof. Set $z = e^{it}$ for $0 \le t \le 2\pi$. Then

$$\cos t = \frac{e^{it} + e^{-it}}{2} = \frac{z^2 + 1}{2z}$$

and $dz = ie^{it}\, dt = iz\, dt$. Hence

$$\int_0^{2\pi} \frac{dt}{\sqrt{5} + \cos t} = \int_{|z|=1} \frac{1}{\sqrt{5} + (z^2 + 1)/(2z)} \frac{dz}{iz}$$

$$= \frac{2}{i} \int_{|z|=1} \frac{dz}{z^2 + 2\sqrt{5}\, z + 1}$$

The only singularities of the integrand are $-\sqrt{5} - 2$ and $-\sqrt{5} + 2$, and of these only $-\sqrt{5} + 2$ lies in the interior of the circle $|z| = 1$ (the singularity $-\sqrt{5} - 2$ lies in the exterior of the circle $|z| = 1$). Thus by Theorems 5.6 and 5.9,

$$\int_0^{2\pi} \frac{dt}{\sqrt{5} + \cos t} = 2\pi i\, \frac{2}{i} \lim_{z \to -\sqrt{5}+2} (z + \sqrt{5} - 2)\frac{1}{z^2 + 2\sqrt{5}\, z + 1}$$

$$= 4\pi^1/_4 = \pi$$

We are next going to determine some methods of computing integrals of the form

$$I = \int_{-\infty}^{\infty} F(x)\, dx$$

where the improper integral I is defined to be

$$\lim_{R \to \infty} \int_{-R}^{R} F(x)\, dx$$

provided that this limit exists. This is usually called the *Cauchy principal value* of the integral, and this definition of integral is different from the stronger definition

$$\lim_{\substack{R \to +\infty \\ r \to -\infty}} \int_r^R F(x)\, dx \qquad (5.8)$$

Clearly if the last integral exists then the Cauchy principal value of it exists and is equal to it, but the converse is not true, as the simple example $F(t) = t^3$ indicates. In the following two theorems the real integrals

$$\int_{-\infty}^{\infty} F(x)\, dx \qquad \int_{-\infty}^{\infty} F(x) \cos mx\, dx \qquad \int_{-\infty}^{\infty} F(x) \sin mx\, dx$$

involved in the conclusions are understood to represent their Cauchy principal value.

Theorem 5.12 Let $z_1, z_2, \ldots, z_n \in \{z \in \mathbf{C} : \operatorname{Im} z > 0\}$, and let F be analytic on $\{z \in \mathbf{C} : \operatorname{Im} z \geq 0\} - \{z_1, \ldots, z_n\}$ with poles at $z_1, z_2, \ldots, z_n$. Suppose further that $\lim_{|z| \to \infty} zF(z) = 0$ in $\{z : \operatorname{Im} z \geq 0\}$. Then

$$\int_{-\infty}^{\infty} F(x)\, dx = 2\pi i \sum_{k=1}^{n} \operatorname{Res}(F, z_k)$$

Proof. Choose R sufficiently large so that all the poles of F lie inside the upper semicircle $\Gamma = \{z \in \mathbf{C} : |z| = R, \operatorname{Im} z \geq 0\}$. See Fig. 5.4. Then by Cauchy's residue theorem, we have

$$2\pi i \sum_{k=1}^{n} \operatorname{Res}(F, z_k) = \int_{-R}^{R} F(x)\, dx + \int_{\Gamma} F(z)\, dz = J_1 + J_2$$

Fig. 5.4

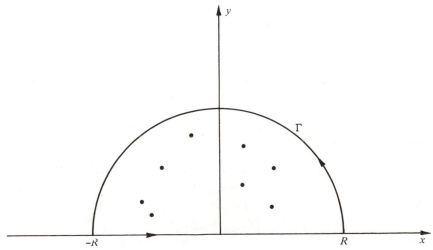

To prove our result it suffices to show that $|J_2| \to 0$ as $R \to +\infty$. In fact, since $\lim_{|z| \to \infty} zF(z) = 0$ in $\{z \in \mathbf{C} : \operatorname{Im} z \geq 0\}$, it follows that given $\varepsilon > 0$, there exists an $R(\varepsilon)$ such that

$$|RF(Re^{i\theta})| < \frac{\varepsilon}{\pi} \quad \text{for } R > R(\varepsilon) \text{ and } \theta \in [0,\pi]$$

Thus if $R > R(\varepsilon)$,

$$|J_2| = \left| \int_0^\pi F(Re^{i\theta}) iRe^{i\theta} \, d\theta \right| < \varepsilon$$

and the proof is complete.

Example 5.12 Show that

$$\int_{-\infty}^{\infty} \frac{dx}{(x^2 + 1)^2 (x^2 + 4)} = \frac{\pi}{9}$$

Proof. The function

$$F(z) = \frac{1}{(z^2 + 1)^2 (z^2 + 4)}$$

clearly satisfies the hypotheses of Theorem 5.11 with $z_1 = i$ and $z_2 = 2i$. Thus

$$\int_{-\infty}^{\infty} \frac{dx}{(x^2 + 1)^2 (x^2 + 4)} = 2\pi i [\operatorname{Res}(F,i) + \operatorname{Res}(F,2i)]$$

But

$$\operatorname{Res}(F,i) = \lim_{z \to i} \frac{d}{dz} (z - i)^2 \frac{1}{(z^2 + 1)^2 (z^2 + 4)} = -\frac{i}{36}$$

and

$$\operatorname{Res}(F,2i) = \lim_{z \to 2i} (z - 2i) \frac{1}{(z^2 + 1)^2 (z^2 + 4)} = -\frac{i}{36}$$

Hence

$$\int_{-\infty}^{\infty} \frac{dx}{(x^2 + 1)^2 (x^2 + 4)} = 2\pi i \left(-\frac{i}{36} - \frac{i}{36} \right) = \frac{\pi}{9}$$

Theorem 5.13 Let F be analytic in $\{z \in \mathbf{C} : \operatorname{Im} z \geq 0\} - \{z_1, \ldots, z_n\}$, where $\operatorname{Im} z_i > 0$ for $i = 1, 2, \ldots, n$, and let $z_1, z_2, \ldots, z_n$ be poles of F. Suppose further that $\lim_{|z| \to \infty} F(z) = 0$ and that $r > 0$. Then

$$\int_{-\infty}^{\infty} F(x) e^{irx} \, dx = 2\pi i \sum_{k=1}^{n} \operatorname{Res}[F(z)e^{irz}, z_k]$$

Proof. Let $\varepsilon > 0$ be given. Choose R sufficiently large so that all the poles of F lie inside the upper semicircle $\Gamma = \{z \in \mathbf{C} : |z| = R \text{ and } \operatorname{Im} z \geq 0\}$. Then by Cauchy's residue theorem, we have

$$2\pi i \sum_{k=1}^{n} \operatorname{Res}(F(z)e^{irz}, z_k) = \int_{-R}^{R} F(x) e^{irx} \, dx + \int_{\Gamma} F(z) e^{irz} \, dz = J_1 + J_2$$

To prove our result it suffices to show that $|J_2| \to 0$ as $R \to +\infty$. In fact if R is so large that $|F(z)| < (r/\pi)\varepsilon$ for $z \in \Gamma$, then since

$$|J_2| = \left| \int_0^\pi F(Re^{i\theta})e^{irRe^{i\theta}} iRe^{i\theta}\, d\theta \right|$$

by using the inequality $\sin\theta \geq 2\theta/\pi$ for $0 \leq \theta \leq \pi/2$, we obtain

$$|J_2| \leq \frac{r}{\pi}\,\varepsilon R \int_0^\pi e^{-rR\sin\theta}\, d\theta$$

$$\leq \frac{2r}{\pi}\,\varepsilon R \int_0^{\pi/2} e^{(-2rR/\pi)\theta}\, d\theta$$

$$= \frac{2r}{\pi}\,\varepsilon R \left(-\frac{\pi}{2rR}\right)(e^{-rR} - 1)$$

$$< \varepsilon(1 - e^{-rR}) < \varepsilon$$

and the proof is complete.

Example 5.13 Show that

$$I_1 = \int_{-\infty}^\infty \frac{\cos 2x}{x^2 + 2x + 2}\, dx = \frac{\pi}{e^2}\cos 2$$

and

$$I_2 = \int_{-\infty}^\infty \frac{\sin 2x}{x^2 + 2x + 2}\, dx = \frac{\pi}{e^2}\sin 2$$

Proof. The function

$$F(z) = \frac{1}{z^2 + 2z + 2}$$

clearly satisfies the conditions of Theorem 5.13 with $z_1 = -1 + i$. Thus

$$\int_{-\infty}^\infty F(x)e^{2ix}\, dx = 2\pi i \operatorname{Res}[F(z)e^{2iz}, -1 + i]$$

But

$$\operatorname{Res}[F(z)e^{2iz}, -1 + i] = \lim_{z \to -1+i} (z + 1 - i)\frac{e^{2iz}}{z^2 + 2z + 2} = \frac{e^{2i(-1+i)}}{2i}$$

Hence

$$\int_{-\infty}^\infty F(x)e^{2ix}\, dx = 2\pi i\,\frac{\cos 2 - i\sin 2}{2ie^2} = \frac{\pi\cos 2 - i\pi\sin 2}{e^2} \qquad (5.9)$$

Also

$$\int_{-\infty}^\infty F(x)e^{2ix}\, dx = I_1 + iI_2 \qquad (5.10)$$

From Eqs. (5.9) and (5.10) the desired results follow.

Finally we use the residue theory to calculate integrals of the form $\int_0^\infty R(x)/x^\alpha\, dx$. Let $\mathscr{A}\mathrm{rg}$ be the "branch of arg" on $\mathbf{C} - \{x: x \geq 0\}$ so that

$0 < \mathscr{A}\text{rg } z < 2\pi$ for all $z \in \mathbf{C} - \{x : x \geq 0\}$, and let $\mathscr{L}\text{og } z = \ln |z| + i\,\mathscr{A}\text{rg } z$ if $z \in \mathbf{C} - \{x : x \geq 0\}$. [In other words, we have merely chosen the branch of log on $\mathbf{C} - \{x : x \geq 0\}$ such that $\mathscr{L}\text{og }(-1) = \pi i$.] Finally if $z \in \mathbf{C} - \{x : x \geq 0\}$, let $z^\alpha = e^{\alpha \mathscr{L}\text{og } z}$. [Thus we see that we have simply chosen the branch of z^α on $\mathbf{C} - \{x : x \geq 0\}$ such that $(-1)^\alpha = e^{\pi i \alpha}$.]

Theorem 5.14 Let R be a rational function with no poles on $\{x : x \geq 0\}$ and let $0 < \alpha < 1$. Then

$$\int_0^\infty \frac{R(x)}{x^\alpha}\, dx$$

converges if and only if $\lim_{x \to \infty} R(x) = 0$. Moreover, if $\lim_{x \to \infty} R(x)$ does equal zero, then

$$\int_0^\infty \frac{R(x)}{x^\alpha}\, dx = \frac{2\pi i}{1 - e^{-2\pi i \alpha}} \sum_{\substack{z_0 \text{ is a} \\ \text{pole of } R}} \text{Res}\left(\frac{R(z)}{z^\alpha}, z_0 \right)$$

Proof. As R is continuous at 0, it is clear that $\int_0^\infty R(x)/x^\alpha\, dx$ exists if and only if $\lim_{x \to \infty} R(x) = 0$. So let us assume that $\lim_{x \to \infty} R(x) = 0$ and let $f : \mathbf{C} - \{x : x \geq 0\} \to \mathbf{C}$ be the analytic function defined by

$$f(z) = \frac{R(z)}{z^\alpha} \qquad \text{if } z \in \mathbf{C} - \{x : x \geq 0\}$$

We are now ready to compute

$$\int_0^\infty \frac{R(x)}{x^\alpha}\, dx$$

Let $0 < \varepsilon < r$. See Fig. 5.5.

Note that $r' \to \infty$ as $r \to \infty$. Setting $\gamma(\varepsilon, r) = \gamma_1(\varepsilon, r) + \gamma_2(r) + \gamma_3(\varepsilon, r) + \gamma_4(\varepsilon)$ (where $\gamma_1, \gamma_2, \gamma_3$, and γ_4 are the curves shown in Fig. 5.5), we have

$$\int_{\gamma(\varepsilon, r)} \frac{R(z)}{z^\alpha}\, dz = \int_{\gamma_1(\varepsilon, r)} f(z)\, dz + \int_{\gamma_2(r)} f(z)\, dz$$
$$+ \int_{\gamma_3(\varepsilon, r)} f(z)\, dz + \int_{\gamma_4(\varepsilon)} f(z)\, dz$$

Let $\Gamma_{(\varepsilon, r')}(t) = t$ for $\varepsilon \leq t \leq r'$. As z^α agrees with $e^{\alpha \,\text{Log } z}$ in the upper half-plane, we see that

$$\int_{\gamma_1(\varepsilon, r)} f(z)\, dz = \int_{\gamma_1(\varepsilon, r)} \frac{R(z)}{e^{\alpha \,\text{Log } z}}\, dz = \int_{\Gamma_{(\varepsilon, r')}} \frac{R(z)}{e^{\alpha \,\text{Log } z}}\, dz = \int_\varepsilon^{r'} \frac{R(x)}{x^\alpha}\, dx$$

Similarly, since z^α agrees with $e^{2\pi i \alpha + \alpha \,\text{Log } z} = e^{2\pi i \alpha} e^{\alpha \,\text{Log } z}$ in the lower half-plane, we have

$$\int_{\gamma_3(\varepsilon, r)} f(z)\, dz = \int_{\gamma_3(\varepsilon, r)} \frac{R(z)}{e^{2\pi i \alpha} e^{\alpha \,\text{Log } z}}\, dz$$
$$= e^{-2\pi i \alpha} \int_{r'}^\varepsilon \frac{R(x)}{x^\alpha}\, dx = -e^{-2\pi i \alpha} \int_\varepsilon^{r'} \frac{R(x)}{x^\alpha}\, dx$$

Fig. 5.5

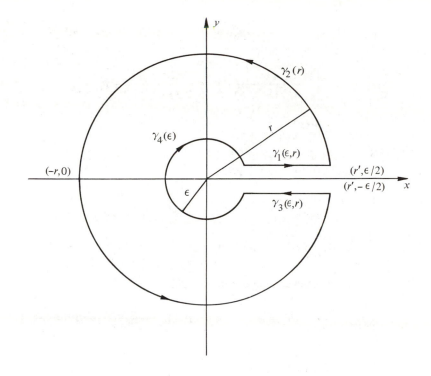

Note that

$$\left| \int_{\gamma_2(R)} \frac{R(z)}{z^\alpha}\, dz \right| \leq \pi r^{1-\alpha} \max_{0 \leq t \leq \pi} R(re^{it}) \to 0 \qquad \text{as } r \to \infty$$

and

$$\left| \int_{\gamma_4(\varepsilon)} \frac{R(z)}{z^\alpha}\, dz \right| \leq \pi \varepsilon^{1-\alpha} \max_{0 \leq t \leq \pi} R(\varepsilon e^{it}) \to 0 \qquad \text{as } \varepsilon \to 0$$

Hence

$$2\pi i \sum_{\substack{z_0 \text{ is a} \\ \text{pole of } R}} \operatorname{Res}\left[\frac{R(z)}{z^\alpha}, z_0 \right] = \lim_{\substack{\varepsilon \to 0 \\ r \to \infty}} \int_{\gamma(\varepsilon,r)} f(z)\, dz$$

$$= (1 - e^{-2\pi i \alpha}) \int_0^\infty \frac{R(x)}{x^\alpha}\, dx$$

Example 5.14 Show that

$$\int_0^\infty \frac{1}{x^{1/2}(1 + x)}\, dx = \pi$$

Proof. Using Theorem 5.14 with $R(x) = 1/(1 + x)$ for $x \geq 0$, we obtain

$$\int_0^\infty \frac{1}{x^{1/2}(1 + x)}\, dx = \frac{2\pi i}{1 - e^{-\pi i}} \operatorname{Res}\left[\frac{1}{z^{1/2}(1 + z)}, -1 \right]$$

Since $-1 = e^{\pi i}$, we have

$$\text{Res}\left[\frac{1}{z^{1/2}(1 + z)}, -1\right] = \lim_{z \to e^{\pi i}} (z + 1)\frac{1}{z^{1/2}(1 + z)} = \frac{1}{e^{\pi i/2}} = \frac{1}{i}$$

Thus

$$\int_0^\infty \frac{1}{x^{1/2}(1 + x)}\,dx = \frac{2\pi i}{1 - e^{-\pi i}}\frac{1}{i} = \pi$$

EXERCISES

5.5.1 Evaluate the following integrals.

a. $\displaystyle\int_0^{2\pi} \frac{\cos 3x}{2 - \cos x}\,dx$ b. $\displaystyle\int_{-\infty}^\infty \frac{8x^6}{(1 + x^4)^2}\,dx$

c. $\displaystyle\int_0^\infty \frac{x \sin x}{x^2 + 9}\,dx$

5.5.2 Show that the following equations hold.

a. $\displaystyle\int_0^\pi \frac{4}{3 + \cos \theta}\,d\theta = \pi\sqrt{2}$ b. $\displaystyle\int_0^\pi \frac{\cos 2\theta}{5 - 4\cos \theta}\,d\theta = \frac{\pi}{12}$

c. $\displaystyle\int_{-\infty}^\infty \frac{1}{(4 + x^2)^2}\,dx = \frac{\pi}{16}$ d. $\displaystyle\int_0^\pi \frac{1}{1 + \sin^2 x}\,dx$

5.5.3 Compute the integrals.

a. $\displaystyle\int_0^\infty \frac{\cos \pi\theta}{\theta^4 + 4}\,d\theta$ b. $\displaystyle\int_0^\infty \frac{\cos \theta}{(1 + \theta^2)^3}\,d\theta$

c. $\displaystyle\int_{-\infty}^\infty \frac{\cos \theta}{(1 + \theta^2)(4 + \theta^2)}\,d\theta$

5.5.4 Show that for $0 < b < a$, the following equations hold.

a. $\displaystyle\int_0^\pi \frac{\sin^4 x}{a + b\cos x}\,dx = \frac{\pi}{b^4}\left[(a^2 - b^2)^{3/2} - a^3 + {}^3/_2 ab^2\right]$

b. $\displaystyle\int_0^{2\pi} \frac{\sin^2 x}{a + b\cos x}\,dx = \frac{2\pi}{b^2}\left(a - \sqrt{a^2 - b^2}\right)$

5.5.5 Evaluate the following integrals.

a. $\displaystyle\int_0^\infty \frac{1}{1 + x^4}\,dx$ b. $\displaystyle\int_0^\infty \frac{x^2}{x^4 + x^2 + 1}\,dx$

c. $\displaystyle\int_0^\infty \frac{1}{x^{2/3}(1 + x)}\,dx$

5.5.6 Show that

a. $\displaystyle\int_0^\infty \frac{x^{\lambda-1}}{1+x}\,dx = \frac{\pi}{\sin \lambda\pi}$ for $0 < \lambda < 1$

b. $\displaystyle\int_0^\infty \frac{x^\lambda}{1+x^2}\,dx = \frac{\pi}{2}\sec\frac{\pi\lambda}{2}$ for $0 < \lambda < 1$

c. $\displaystyle\int_0^\pi \sin^{2n} x\,dx = \frac{(2n)!}{(2^n n!)^2}$ for $n = 1, 2, 3, \ldots$

d. $\displaystyle\int_0^\pi \frac{\cos 2x}{a^2 - 2a\cos x + 1}\,dx = \frac{\pi a^2}{1 - a^2}$ for $-1 < a < 1$

e. $\displaystyle\int_0^\infty \frac{dx}{(x^2 + a^2)(x^2 + b^2)} = \frac{\pi}{2ab(a+b)}$ for $a > 0$ and $b > 0$

5.6
THE ARGUMENT PRINCIPLE AND ROUCHÉ'S THEOREM

This section deals with the number of zeros and poles of a meromorphic function.

Theorem 5.15 Let f be meromorphic on D and let $\gamma: [a,b] \to \operatorname{dmn} f$ be a closed curve which is contractible in D such that $f(\gamma(t)) \neq 0$ for all $t \in [a,b]$. Finally, let

$$\mathscr{P} = \{z \in D: I(\gamma,z) \neq 0 \text{ and } \mathcal{O}(f,z) \neq 0\}$$

Then $\mathscr{P}$ is a finite set, and

$$\int_\gamma \frac{f'(z)}{f(z)}\,dz = 2\pi i \sum_{z \in \mathscr{P}} I(\gamma,z)\mathcal{O}(f,z)$$

Proof. By applying Corollary 4.14 and Theorem 5.3 to f and $1/f$, we see that $\mathscr{P}$ is a finite set. Recall that if $z_0 \in D$ and $m = \mathcal{O}(f,z_0)$, then by property 2 in Corollary 5.3 there exists a function ϕ analytic at z_0 with $\phi(z_0) \neq 0$ such

---

EUGÈNE ROUCHÉ

Eugène Rouché was born in Somières, France, on August 18, 1832. He worked on complex functions, descriptive geometry, algebra, and probability theory. Unfortunately, the details of his life are unknown to us. He died in Lunel, France, on August 19, 1910.

that if z is near z_0, then $f(z) = (z - z_0)^m \phi(z)$. Thus it follows that $f'(z) = m(z - z_0)^{m-1} \phi(z) + (z - z_0)^m \phi'(z)$, and so

$$\frac{f'(z)}{f(z)} = \frac{m}{z - z_0} + \frac{\phi(z)}{\phi(z)} \qquad \text{for } z \text{ near } z_0$$

Hence Res $(f'/f, z_0) = m = \mathcal{O}(f,z_0)$ which (together with Theorem 5.10) proves the theorem.

If f is an analytic function and $w \in \mathbf{C}$, we shall let $f - w$ be the function $(f - w)(z) = f(z) - w$. Thus if $z_0 \in$ dmn f, then $\mathcal{O}(f - w, z_0)$ is just the multiplicity of z_0 as a root of the equation $f(z) = w$, whereas if z_0 is a pole of f of order m, then z_0 is also a pole of $f - w$ of order m and so in this case $\mathcal{O}(f - w, z_0) = -m$. Note in particular that if $w = 0$, then $\mathcal{O}(f - 0, z_0) = \mathcal{O}(f,z_0)$.

Corollary 5.4 Let f be meromorphic on D, let $w \in \mathbf{C}$, and let $\gamma: [a,b] \rightarrow$ dmn f be a closed curve which is contractible in D such that $f(\gamma(t)) \neq w$ for all $t \in [a,b]$. Finally, let

$$\mathcal{P} = \{z \in D: I(\gamma,z) \neq 0 \text{ and } \mathcal{O}(f - w, z) \neq 0\}$$

Then $\mathcal{P}$ is a finite set, and

$$\int_\gamma \frac{f'(z)}{f(z) - w} \, dz = 2\pi i \sum_{z \in \mathcal{P}} I(\gamma,z) \mathcal{O}(f - w, z)$$

The following corollary provides us with a computational technique often useful in determining the number of zeros and poles of an analytic function (with the multiplicities of zeros and poles being counted).

Corollary 5.5. *The Argument Principle* Let f be meromorphic on D and let $\gamma: [a,b] \rightarrow$ dmn f be a positively oriented simple closed curve which is contractible in D such that $f(\gamma(t)) \neq 0$ for all $t \in [a,b]$. Finally, let

$$\mathcal{P} = \{z \in D: z \text{ is in the interior of } \gamma \text{ and } \mathcal{O}(f,z) \neq 0\}$$

Then $\mathcal{P}$ is a finite set, and

$$\frac{1}{2\pi i} \int_\gamma \frac{f'(z)}{f(z)} \, dz = \sum_{z \in \mathcal{P}} \mathcal{O}(f,z)$$

that is,

$$\frac{1}{2\pi i} \int_\gamma \frac{f'(z)}{f(z)} \, dz$$

is the number of zeros of f inside γ minus the number of poles of f inside γ, where the sum is weighted according to the multiplicity of the zero or pole.

Corollary 5.6 Let f be meromorphic on D, let $w \in \mathbf{C}$, let $\gamma: [a,b] \rightarrow$ dmn f be a simple closed curve which is contractible in D such that $f(\gamma(t)) \neq w$ for all

$t \in [a,b]$, and let $\mathscr{P} = \{z \in D : z$ is in the interior of γ and $\mathcal{O}(f - w, z) \neq 0\}$. Then $\mathscr{P}$ is a finite set, and

$$\frac{1}{2\pi i} \int_\gamma \frac{f'(z)}{f(z) - w} \, dz = \sum_{z \in \mathscr{P}} \mathcal{O}(f - w, z)$$

Example 5.15 Compute

$$\int_{|z|=2} \frac{z^5 - 1}{z^6 - 6z + 5} \, dz$$

Solution. First observe that if $|z| \geq 2$, then $|z^6 - 6z + 5| \geq |z|^6 - 6|z| - 5 \geq |z|(|z|^5 - 6) - 5 \geq 2(32 - 6) - 5 = 47$, and so $z^6 - 6z + 5 \neq 0$ if $|z| \geq 2$. Hence the zeros of $z^6 - 6z + 5$ lie inside the circle $|z| = 2$.

Set $f(z) = z^6 - 6z + 5$ for $z \in \mathbf{C}$. Then by Corollary 5.5, it follows (as f has no poles) that

$$\frac{1}{2\pi i} \int_{|z|=2} \frac{6z^5 - 6}{z^6 - 6z + 5} \, dz = \frac{1}{2\pi i} \int_{|z|=2} \frac{f'(z)}{f(z)} \, dz = 6$$

and so

$$\int_{|z|=2} \frac{z^5 - 1}{z^6 - 6z + 5} \, dz = 2\pi i$$

Example 5.16 Show that

$$\int_{|z|=4} \frac{(z - 1) \cot z - 1}{z - 1} \, dz = 2\pi i$$

Solution. Let

$$f(z) = \frac{\sin z}{z - 1} \qquad \text{if } z \in \mathbf{C} - \{0\}$$

Then $f : \mathbf{C} - \{1\} \to \mathbf{C}$ is analytic, and

$$\frac{f'(z)}{f(z)} = \frac{(z - 1) \cot z - 1}{z - 1}$$

Note that the zeros of f inside $|z| = 4$ are $-\pi$ and π and $\mathcal{O}(f, -\pi) = 1 = \mathcal{O}(f, \pi)$, while the only pole of f inside $|z| = 4$ is the simple pole 1. Thus

$$\int_{|z|=4} \frac{(z - 1) \cot z - 1}{z - 1} \, dz = 2\pi i [\mathcal{O}(f, -\pi) + \mathcal{O}(f, 0) + \mathcal{O}(f, \pi)]$$
$$= 2\pi i (1 - 1 + 1) = 2\pi i$$

Corollary 5.7 Let $f : D \to \mathbf{C}$ be analytic, let $w \in \mathbf{C}$, let $\gamma : [a,b] \to D$ be a simple closed curve which is contractible in D such that $f(\gamma(t)) \neq w$ for all $t \in [a,b]$, and let $\mathscr{P} = \{z : z$ is in the interior of γ and $f(z) = w\}$. Then $\mathscr{P}$ is a finite set and

$$\frac{1}{2\pi i} \int_\gamma \frac{f'(z)}{f(z) - w} \, dz = \sum_{z \in \mathscr{P}} \mathcal{O}(f - w, z)$$

Theorem 5.16 Let f be analytic at z_0 and let $\mathcal{O}(f - f(z_0), z_0) = k > 0$ [in other words, z_0 is a root of order k of the equation $f(z) = f(z_0)$]. Then there exists $R > 0$ such that if $0 < |v - z_0| < R$, then $f(z_0) \neq f(v)$, and moreover, $\{z: |z - z_0| < R$ and $f(z) = f(v)\}$ has k distinct elements.

Proof. As f is nonconstant near z_0, there exists $R_1 > 0$ such that if $0 < |z - z_0| \leq R_1$, then $f(z) \neq f(z_0)$ and $f'(z) \neq 0$. We also see that there exists $R > 0$ with $R_1 \geq R$ such that if $|v - z_0| < R$ and $|z - z_0| = R_1$, then $f(v) \neq f(z)$. Let $\gamma(t) = z_0 + R_1 e^{it}$ for $0 \leq t \leq 2\pi$ and let $0 < |v - z_0| < R$. Then if $\mathcal{P}$ is as in Corollary 5.7 [with $w = f(v)$],

$$\sum_{z \in \mathcal{P}} \mathcal{O}(f - f(v), z) = \frac{1}{2\pi i} \int_\gamma \frac{f'(z)}{f(z) - f(v)} \, dz$$

$$= I[f \circ \gamma, f(v)]$$

$$= I[f \circ \gamma, f(z_0)] \qquad \text{(by Theorem 4.17)}$$

$$= \frac{1}{2\pi i} \int_\gamma \frac{f'(z)}{f(z) - f(z_0)} \, dz$$

$$= \mathcal{O}(f, z_0) = k$$

So because $\mathcal{O}(f - f(v), z) = 1$ for each $z \in \mathcal{P}$, the result follows.

The next result is due to Eugène Rouché (1832–1910, French).

Theorem 5.17. *Rouché's Theorem* Let f and g be analytic on a simply connected domain D, and let $\gamma: [a,b] \to D$ be a simple closed curve. Assume that $|f(\gamma(t))| > |g(\gamma(t))|$ for all $t \in [a,b]$. Then f and $f + g$ have the same number of zeros in the interior of γ (where the multiplicities of the zeros have been counted).

Proof. From $|f(\gamma(t))| > |g(\gamma(t))|$ for all $t \in [a,b]$, it follows that the functions f and $f + g$ do not vanish on γ. Since f and $f + g$ are analytic in D, in view of the argument principle, it suffices to show that

$$\int_\gamma \frac{f'(z) + g'(z)}{f(z) + g(z)} \, dz = \int_\gamma \frac{f'(z)}{f(z)} \, dz$$

In fact,

$$\int_\gamma \left[\frac{f'(z) + g'(z)}{f(z) + g(z)} - \frac{f'(z)}{f(z)} \right] dz = \int_\gamma \frac{g'(z)f(z) - g(z)f'(z)}{[f(z) + g(z)]f(z)} \, dz$$

Setting $h = g/f$, the above integral takes the form

$$\int_\gamma \frac{h'(z)f^2(z)}{[f(z) + g(z)]f(z)} \, dz = \int_\gamma \frac{h'(z)}{1 + h(z)} \, dz = 0$$

and the proof is complete.

Example 5.17 Show that the three roots of the cubic equation $z^3 - 6z + 8 = 0$ lie between the circles $|z| = 1$ and $|z| = 3$.

Proof. Consider the functions $g_1(z) = z^3 - 6z$ and $f_1(z) = 8$ for $z \in \mathbf{C}$. Clearly on $|z| = 1$, we have

$$|g_1(z)| = |z^3 - 6z| \leq |z|^3 + 6|z| = 7 < 8 = f_1(z)$$

Thus by Rouché's theorem the functions f_1 and $f_1 + g_1$ have the same number of zeros inside the circle $|z| = 1$. Since $f_1(z) = 8$ has no zero inside $|z| = 1$, it follows that the three roots of $f_1 + g_1$ lie outside the circle $|z| = 1$. Next consider the functions $f_2(z) = z^3$ and $g_2(z) = -6z + 8$ for $z \in \mathbf{C}$. Clearly on $|z| = 3$, we have

$$|g_2(z)| = |-6z + 8| \leq 6|z| + 8 = 26 < 27 = |f_2(z)|$$

Thus by Rouché's theorem the functions f_2 and $f_2 + g_2$ have the same number of zeros inside the circle $|z| = 3$. Since $f_2(z) = z^3$ has "three" zeros inside the circle $|z| = 3$, it follows that the three zeros of $f_2 + g_2$ lie inside the circle $|z| = 3$. Combining the two conclusions of this proof, the desired result is established.

EXERCISES

5.6.1 Compute the following integrals.

a. $\displaystyle\int_{|z|=3} \frac{z^2}{z^3 - 2}\, dz$ b. $\displaystyle\int_{|z|=2} \frac{2z^3 + 1}{z^4 + 2z + 1}\, dz$

5.6.2 Show that the equation $e^z - 3z^7 = 0$ has seven zeros in the interior of the circle $|z| = 1$.

5.6.3 Find the number of zeros of the equation $e^2 - ze^z = 0$ in the interior of the circle $|z| = 1$.

5.6.4 Use Rouché's theorem to prove the fundamental theorem of algebra.

Chapter **6**   Applications

6.1

INTRODUCTION

In this chapter we present several applications of complex function theory to physics, mathematics, and engineering. This chapter is by no means meant to be a complete and self-contained treatment of applications. Rather, its purpose is to indicate to the student the broad-ranging and diverse nature of the types of applications possible. In Sec. 6.2 we present an application of complex variables to two-dimensional flows of fluids. Next we apply the theory of complex power series to solve ordinary differential equations of physical interest. In Secs. 6.4 and 6.5 we state some facts about Fourier and Laplace transforms and use them to present two examples (one electrical and one mechanical) that show the usefulness of complex variables in the solution of interesting physical problems.

6.2

TWO-DIMENSIONAL FLOWS

In this section, we present a connection between the theory of analytic functions and that of two-dimensional flows of ideal fluids. An *ideal fluid* is a fluid which is *incompressible* (has constant density) and *frictionless* (has zero viscosity). A *two-dimensional flow* (or a *velocity field*) over a domain $D \subset \mathbf{R}^2$ is a continuous function $V: D \to \mathbf{R}^2$. If $V(x,y) = (u(x,y), v(x,y))$ for all $(x,y) \in D$, then $u(x,y)$ and $v(x,y)$ can be physically interpreted as the first and second components, respectively, of the velocity $V(x,y)$ of the fluid at the point (x,y). Since dmn $V \subset \mathbf{R}^2$, we see that we are also assuming that the flow is *stationary* [that is, $V(x,y)$ depends only on the position (x,y) and is independent of time]. A streamline of the flow V is a curve γ in D whose tangent at any point (x,y) on the curve is the vector $V(x,y) = (u(x,y), v(x,y))$; in other words, the tangent vector $\gamma'(t)$ to γ at $\gamma(t)$ should satisfy $\gamma'(t) = V(\gamma(t))$ for all $t \in$ dmn γ. Hence we see that physically the streamlines represent the lines of motion of the flow. A curve δ in D is called an *equipotential line* to the flow V if it is orthogonal to all streamlines of the flow V [that is, if γ is a streamline in D of V and $\delta(s) = \gamma(t)$ for some s in the domain of δ and some t in the domain of γ, then $\delta'(s) \cdot \gamma'(t) = 0$ where $\cdot$ is the familiar dot product of two vectors].

If $F: D \to \mathbf{R}$ and $G: D \to \mathbf{R}$ are continuous functions, and if $\gamma: [a,b] \to D$ is a contour given by $\gamma(t) = (x(t), y(t))$ for $a \le t \le b$, then we define

$$\int_\gamma F \, dx + G \, dy = \int_a^b [F(\gamma(t))x'(t) + G(\gamma(t))y'(t)] \, dt$$

We say that a flow $V = (u,v)$ over a domain D is *divergence-free* if for every closed contour γ in D,

$$\int_\gamma u \, dy - v \, dx = 0 \tag{6.1}$$

The integral $\int_\gamma u \, dy - v \, dx$ is called the *flux* of the fluid over γ and, as we shall see in Eq. (6.6), represents the amount of fluid going outside of γ minus the amount of fluid coming inside γ.

We say that a flow $V = (u,v)$ over a domain D is *circulation-free* (or *irrotational*) if for every closed contour γ in D

$$\int_\gamma u \, dx + v \, dy = 0 \tag{6.2}$$

The integral $\int_\gamma u \, dx + v \, dy$ is called the *circulation* of the fluid over γ.

We can now prove the following theorem.

Theorem 6.1 Let $V = (u,v)$ be a two-dimensional flow over a simply connected domain D. Then V is divergence-free and circulation-free if and only if the function $f = u - iv$ is analytic in D.

Proof. If $V = (u,v)$ is divergence-free and circulation-free, then for any triangle γ which together with its interior is contained in D we have, in view of Eqs. (6.1) and (6.2),

$$\int_\gamma f(z) \, dz = \int_\gamma (u - iv)(dx + i \, dy)$$

$$= \int_\gamma u \, dx + v \, dy + i \int_\gamma u \, dy - v \, dx = 0 \tag{6.3}$$

Thus by Morera's Theorem 4.9, the function $f = u - iv$ is analytic in D. Conversely, if $f = u - iv$ is analytic in D, then for every closed curve γ in D we have

$$0 = \int_\gamma f(z)\,dz = \int_\gamma u\,dx + v\,dy + i \int_\gamma u\,dy - v\,dx$$

which implies Eqs. (6.1) and (6.2), and the proof is complete.

For the rest of this section, we shall assume that $V = (u,v)$ is a two-dimensional, divergence-free, and circulation-free flow over a simply connected domain D; then the analytic function $f = u - iv$ is called the *complex velocity* of the flow, and (after we fix $z_0 \in D$) the analytic function

$$P(z) = \int_{z_0}^z f(\zeta)\,d\zeta \qquad \text{for } z \in D \tag{6.4}$$

is called the *complex potential* of the flow (where P is well defined by Corollary 4.10b). Note that up to the choice of an additive constant (determined by the choice of z_0), P is uniquely determined by f. If for $z = (x,y) \in D$, we write $P(z) = U(x,y) + iW(x,y)$, it follows from Eq. (6.4) as in the proof of Theorem 6.1 that

$$U(x,y) = \int_{(x_0,y_0)}^{(x,y)} u\,dx + v\,dy$$

and

$$W(x,y) = \int_{(x_0,y_0)}^{(x,y)} u\,dy - v\,dx$$

(6.5)

Let $\gamma: [a,b] \to D$ be a contour in D that has been parametrized with respect to arclength. In other words if $\gamma(t) = (x(t),y(t))$ for all $t \in [a,b]$, assume that $|\gamma'(t)| = \sqrt{[x'(t)]^2 + [y'(t)]^2} = 1$ for all $t \in [a,b]$. Let $V_T(t)$ and $V_N(t)$ be the signed lengths of the projections of $V(\gamma(t))$ to the tangent and outward normal of γ at $\gamma(t) = (x,y)$, respectively. See Fig. 6.1. Recall that $V_T(t) = V(\gamma(t)) \cdot \gamma'(t)$, while if we let $N(t)$ be the unit outward normal to γ at $\gamma(t)$ [recall that $N(t) = -i\gamma'(t) = (y'(t), -x'(t))$], then $V_N(t) = V(\gamma(t)) \cdot N(t)$. Hence $V_T(t) = u(\gamma(t))x'(t) + v(\gamma(t))y'(t)$, whereas $V_N(t) = u(\gamma(t))y'(t) - v(\gamma(t))x'(t)$. Using these relations in Eq. (6.5), after possibly modifying U and W by an additive constant in order to assume $(x_0,y_0) = \gamma(a)$, we have that if (x,y) lies on γ,

$$W(x,y) = \int_{(x_0,y_0)}^{(x,y)} V_N(t)\,dt$$

and

$$U(x,y) = \int_{(x_0,y_0)}^{(x,y)} V_T(t)\,dt$$

(6.6)

Fig. 6.1

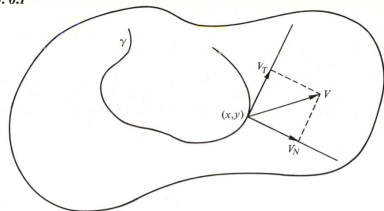

where the above integration is merely integrating along γ from (x_0, y_0) to (x, y). Clearly γ is a streamline of the flow V if and only if $V_N(t) = 0$ for all $t \in [a, b]$, whereas γ is an equipotential line of the flow V if and only if $V_T(t) = 0$ for all $t \in [a, b]$. With this in mind, we easily obtain the following result.

Corollary 6.1 Let V be a two-dimensional, divergence-free, and circulation-free flow over a simply connected domain D. Let γ be a contour in D. Then

a. γ is a streamline if and only if $W(x, y) = $ constant on γ
b. γ is an equipotential line if and only if $U(x, y) = $ constant on γ

Example 6.1 Determine the streamlines and the equipotential lines of the flow over $\mathbf{C}$ with complex potential $P(z) = 2z + i$. What is the flow in this case?

Solution. Here we have

$$U + iW = P(z) = 2x + i(2y + 1)$$

Thus the streamlines are the curves $2y + 1 = c$ for some $c \in \mathbf{C}$; that is, they are straight lines parallel to the x axis. The equipotential lines are the curves $2x = c$ for some $c \in \mathbf{C}$; that is, they are straight lines parallel to the y axis. If $V = (u, v)$, then $u - iv = f(z) = P'(z) = 2$, and so $V = (u, v) = (2, 0)$.

6.2.1 Determine the streamlines, equipotential lines, and modulus of the complex velocity of the flow over $\mathbf{C}$ with complex potential given by
(a) $P(z) = -iz + 3 + 5i$; (b) $P(z) = iz^2 - 1$.
6.2.2 Prove Corollary 6.1.

After using some quite reasonable approxima-
tions, differential equations occur very naturally as mathematical models of
various physical phenomena.

Differential equations occur, for example, in the theory of electric circuits,
in continuous mechanics, in quantum theory, and in the quantum-mechanical
theory of potential scattering. By means of some standard techniques, most
of these differential equations can be reduced to a *second-order linear ordinary
differential equation* of the form

$$w'' + p(z)w' + q(z)w = 0 \tag{6.7}$$

where p and q are analytic functions in a (possibly deleted) neighborhood of
the origin.

A function w which satisfies Eq. (6.7) in a domain D is called a *solution*
of (6.7) in D. Two solutions w_1 and w_2 of Eq. (6.7) in a domain D are called
linearly dependent in D if there exist constants $\lambda_1, \lambda_2 \in \mathbf{C}$ not both zero such
that

$$\lambda_1 w_1(z) + \lambda_2 w_2(z) = 0$$

for all $z \in D$. Otherwise the two solutions are called *linearly independent*. If
w_1 and w_2 are two linearly independent solutions of Eq. (6.7) then any
solution of (6.7) is of the form $\lambda_1 w_1 + \lambda_2 w_2$ for some $\lambda_1, \lambda_2 \in \mathbf{C}$. This basic
fact is proved in courses about equations of the form of (6.7) and will be
taken for granted here.

When the coefficients p and q of Eq. (6.7) are analytic functions in a
neighborhood of zero and therefore have Taylor-series expansions about
zero with radius $R > 0$, it can be shown by direct substitution that (6.7) has
two linearly independent solutions w_1 and w_2 having Taylor-series expansions about zero with radius of convergence equal to R. That is,

$$w_1(z) = \sum_{n=0}^{\infty} a_n z^n \quad \text{and} \quad w_2(z) = \sum_{n=0}^{\infty} b_n z^n \quad \text{for } |z| < R$$

Example 6.2 Find two linearly independent solutions of the differential
equation

$$w'' + zw' + w = 0 \tag{6.8}$$

Solution. Here $p(z) = z$, $q(z) = 1$, and $R = \infty$. Searching for a solution of
Eq. (6.8) of the form

$$w(z) = \sum_{n=0}^{\infty} a_n z^n$$

we obtain, in view of Eq. (6.8) and Theorem 3.1,

$$\sum_{n=2}^{\infty} n(n-1)a_n z^{n-2} + z \sum_{n=1}^{\infty} na_n z^{n-1} + \sum_{n=0}^{\infty} a_n z^n = 0$$

Thus

$$\sum_{n=0}^{\infty} [(n+2)(n+1)a_{n+2} + na_n + a_n]z^n = 0$$

By Corollary 3.3 we obtain the *recurrence relation*

$$a_{n+2} = -\frac{1}{n+2} a_n \qquad n = 0, 1, 2, \ldots$$

Hence $a_2 = -(^1/_2)a_0$, $a_3 = -(^1/_3)a_1$, $a_4 = (1/2 \cdot 4)a_0$, $a_5 = (1/3 \cdot 5)a_1, \ldots$ and consequently

$$w(z) = a_0 \left[1 + \sum_{n=1}^{\infty} \frac{(-1)^n z^{2n}}{2 \cdot 4 \cdots (2n)} \right] + a_1 \left[z + \sum_{n=1}^{\infty} \frac{(-1)^n z^{2n+1}}{3 \cdot 5 \cdots (2n+1)} \right]$$

It follows that

$$w_1(z) = 1 + \sum_{n=1}^{\infty} \frac{(-1)^n z^{2n}}{2 \cdot 4 \cdots (2n)}$$

and

$$w_2(z) = z + \sum_{n=1}^{\infty} \frac{(-1)^n z^{2n+1}}{3 \cdot 5 \cdots (2n+1)}$$

are two linearly independent solutions of Eq. (6.8) having radius of convergence $R = \infty$.

The method of power series also offers a powerful technique of finding the two linearly independent solutions of Eq. (6.7) when the coefficients p and q of (6.7) are of the form

$$p(z) = \frac{p_{-1}}{z} + p_0 + p_1 z + p_2 z^2 + \cdots \tag{6.9}$$

and

$$q(z) = \frac{q_{-2}}{z^2} + \frac{q_{-1}}{z} + q_0 + q_1 z + q_2 z^2 + \cdots \tag{6.10}$$

in a deleted neighborhood $N_R(0) - \{0\}$ of the origin. When the coefficients of Eq. (6.7) are given by Eqs. (6.9) and (6.10) with $|p_{-1}| + |q_{-2}| \neq 0$, we say that 0 is a singularity of (6.7) of the *first kind*.

We shall now restrict ourselves until the end of this section to the study of equations of type (6.7) that have a singularity of the first kind at the origin;

our justification is that these equations occur frequently in physical applications. A well-studied example of this type of equation is Bessel's differential equation (Friedrich Wilhelm Bessel, 1784–1846, German) of order p:

$$w'' + \frac{1}{z} w' + \left(1 - \frac{p^2}{z^2}\right) w = 0 \qquad (6.11)$$

Searching for a solution of Eq. (6.7) of the form

$$w(z) = z^\lambda \sum_{n=0}^{\infty} a_n z^n$$

(where $z^\lambda = e^{\lambda L(z)}$ for some branch L of log), we first observe (applying the method of Example 6.2) that λ must satisfy the quadratic equation

$$\lambda(\lambda + p_{-1} - 1) + q_{-2} = 0 \qquad (6.12)$$

which is called the *indicial equation* of Eq. (6.7). Let λ_1 and λ_2 be the two roots of the indicial equation (6.12). The following facts can be proved.

1. If $\lambda_1 - \lambda_2$ is not an integer, then Eq. (6.7) has two linearly independent solutions near zero (where by *near* we mean *on the domain of L*) of the form

$$w_1(z) = z^{\lambda_1} \sum_{n=0}^{\infty} a_n z^n \quad \text{and} \quad w_2(z) = z^{\lambda_2} \sum_{n=0}^{\infty} b_n z^n$$

with $a_0 b_0 \neq 0$.

2. If $\lambda_1 = \lambda_2$, then Eq. (6.7) has two linearly independent solutions near zero of the form

$$w_1(z) = z^{\lambda_1} \sum_{n=0}^{\infty} a_n z^n$$

and

$$w_2(z) = z^{\lambda_1} \left(L(z) \sum_{n=0}^{\infty} a_n z^n + \sum_{n=0}^{\infty} b_n z^n \right)$$

3. If $\lambda_1 = \lambda_2 + m$, where m is a positive integer, then Eq. (6.7) has two linearly independent solutions near zero of the form

$$w_1(z) = z^{\lambda_1} \sum_{n=0}^{\infty} a_n z^n$$

and

$$w_2(z) = c z^{\lambda_1} L(z) \sum_{n=0}^{\infty} a_n z^n + z^{\lambda_2} \sum_{n=0}^{\infty} b_n z^n$$

Example 6.3 Find a solution of Bessel's Equation (6.11).

Solution. The indicial equation of (6.11) is

$$\lambda(\lambda + 1 - 1) - p^2 = 0$$

Thus $\lambda_1 = p$ and $\lambda_2 = -p$. Assume that the roots have been ordered so that Re $\lambda_1 \geq 0$. From the above mentioned facts 1, 2, and 3, we see that Eq. (6.11) must have a solution of the form

$$w_1(z) = z^p \sum_{n=0}^{\infty} a_n z^n$$

Substituting $w_1(z)$ in (6.11) we find after several manipulations that

$$w_1(z) = \left(\frac{z}{2}\right)^p \left[1 - \frac{1}{p+1}\left(\frac{z}{2}\right)^2 + \frac{1}{2!\,(p+1)(p+2)}\left(\frac{z}{2}\right)^4 + \cdots\right]$$

$$= \left(\frac{z}{2}\right)^p \sum_{n=0}^{\infty} \frac{(-1)^n}{n!\,(n+p)!}\left(\frac{z}{2}\right)^{2n}$$

The solution w_1 computed in Example 6.3 is called the *Bessel function of the first kind of order p* and is denoted by J_p. Thus

$$J_p(z) = \left(\frac{z}{2}\right)^p \sum_{n=0}^{\infty} \frac{(-1)^n}{n!\,(n+p)!}\left(\frac{z}{2}\right)^{2n}$$

There is voluminous literature about the properties of Bessel functions which the interested reader may investigate.

EXERCISES

6.3.1 Find two linearly independent solutions of the differential equations

$$w'' + zw' - w = 0$$

6.3.2 Find power-series solutions of the differential equations.

  a. $z^2 w'' + zw' + (z^2 - 1)w = 0$

  b. $w'' - \dfrac{1}{2z} w' + \left(\dfrac{1}{2z} - \dfrac{1}{5z^2}\right) w = 0$

*6.4

FOURIER TRANSFORMS

Let $f: \mathbf{R} \to \mathbf{C}$ be a piecewise-continuous differentiable function which is absolutely integrable on $\mathbf{R}$ [that is, $\int_{-\infty}^{\infty} |f(t)|\,dt < \infty$]. The function F given by

$$F(\omega) = \mathscr{F}[f](\omega) = \int_{-\infty}^{\infty} f(t)e^{-i\omega t}\,dt \qquad (6.13)$$

* Sections 6.4 and 6.5 were contributed by Prof. A. Poularikas.

is then well defined by each $\omega \in \mathbf{R}$ and is called the *Fourier transform* of f (Jean Fourier, 1768–1830, French). In addition, it can be shown that f and F are related by the formula

$$\frac{f(t^+) + f(t^-)}{2} = \frac{1}{2\pi} \int_{-\infty}^{\infty} F(\omega)e^{i\omega t}\, d\omega \qquad (6.14)$$

for each $t \in \mathbf{R}$, where

$$f(t^+) = \lim_{h \to 0^+} f(t + h) \qquad \text{and} \qquad f(t^-) = \lim_{h \to 0^-} f(t + h)$$

and where the integral in Eq. (6.14) is interpreted as a Cauchy principal value. In particular, if f is continuous at t, then (6.14) becomes

$$f(t) = \frac{1}{2\pi} \int_{-\infty}^{\infty} F(\omega)e^{i\omega t}\, d\omega \qquad (6.14a)$$

JEAN BAPTISTE JOSEPH FOURIER

Jean Baptiste Joseph Fourier was born at Auxerre, France, on March 21, 1768, the son of a tailor. Both his parents died before he was nine years old. Thanks to the charitable Madame Moiton and to the Bishop of Auxerre, the orphaned boy was sent to the town's military school. From 1787 to 1794, Fourier taught in secondary schools and also became actively involved in the French Revolution. As a result of this latter activity, Fourier spent some time in the prison of Auxerre in 1794. After his imprisonment, Fourier taught mathematics at the École Normale and the École Polytechnique and did research in the theory of equations and applied mathematics.

In 1798 Fourier was chosen to accompany Napoleon on his expedition to Egypt. A year later, Napoleon appointed him leader of a scientific expedition to investigate the monuments and inscriptions in Upper Egypt. In 1801 Fourier returned to France, where in spite of the heavy administrative duties assigned to him by Napoleon (who in return made him baron in 1808), he nevertheless found time to continue his research on the mathematical theory of heat. His most important work, *Théorie Analytique de la Chaleur*, appeared in 1822.

Fourier died in Paris on May 16, 1830, at the age of 62.

The integral in the right-hand side of Eq. (6.14) is denoted by $\mathscr{F}^{-1}[F(\omega)]$ and called the *inverse Fourier transform* of F. Equations (6.13) and (6.14) are known as the *Fourier transform pair*.

Let f, f_1, f_2, and $g^{(n)}$ be piecewise-continuous differentiable functions from **R** into **C** which are absolutely integrable on **R**. From the definition of the Fourier transform, the following properties follow.

1. *Linearity.* For any complex numbers a_1 and a_2,

$$\mathscr{F}[a_1 f_1 + a_2 f_2] = a_1 \mathscr{F}[f_1] + a_2 \mathscr{F}[f_2]$$

2. *Time shifting.* $\mathscr{F}[f(t - t_0)](\omega) = e^{-it_0\omega}\mathscr{F}[f](\omega)$

3. *Uniqueness.* If f_1 and f_2 have the same Fourier transform, then $f_1(t) = f_2(t)$ for all $t \in \mathbf{R}$ at which f_1 and f_2 are both continuous.

4. $\mathscr{F}[g^{(n)}](\omega) = (i\omega)^n \mathscr{F}[g](\omega)$.

The following example indicates the connection between complex variables and Fourier transforms, and shows the usefulness of the latter in solving electrical problems.

Example 6.4 Find the charge q flowing in the circuit of Fig. 6.2 when a time-varying voltage source v is applied to it. In particular, find q when $v(t) = \delta(t)$, where δ is the Dirac delta function, $R = 4\,\Omega$, $L = 2$ H, and $C = 0.25$ F.

Solution. Figure 6.2 represents a circuit with resistance R measured in ohms (Ω), inductance L in henrys (H), capacitance C in farads (F), and a time-varying voltage v. Applying Kirchhoff's voltage law ("the algebraic sum of voltages around a closed circuit equals zero"), we obtain the integro-differential equation

$$v(t) - RI - L\frac{dI}{dt} - \frac{1}{C}\int_{t_0}^{t} I(\xi)\,d\xi = 0 \qquad (6.15)$$

where $I = dq/dt = q'$ is the current flowing in the circuit with the initial condition $q(t_0) = 0$ [here I is being measured in coulombs per second

Fig. 6.2

(C/sec) and q in coulombs]. By setting $I = q'$ in Eq. (6.15) we obtain, after some elementary manipulations,

$$Lq''(t) + Rq'(t) + \frac{1}{C} q(t) = v(t) \tag{6.16}$$

After taking the Fourier transform of both sides of Eq. (6.16) and setting $\mathscr{F}[v](\omega) = V(\omega)$, we obtain

$$\left[(i\omega)^2 + \frac{R}{L} (i\omega) + \frac{1}{LC} \right] \mathscr{F}[q](\omega) = \frac{1}{L} V(\omega)$$

By solving with respect to $\mathscr{F}[q](\omega)$, we get

$$\mathscr{F}[q](\omega) = \frac{1}{L} \frac{V(\omega)}{1/(LC) - \omega^2 + i(R/L)\omega}$$

and so

$$q(t) = \frac{1}{2\pi L} \int_{-\infty}^{\infty} \frac{V(\omega)e^{i\omega t}}{1/(LC) - \omega^2 + i(R/L)\omega} \, d\omega \tag{6.17}$$

In particular, when $R = 4\,\Omega$, $L = 2$ H, $C = 0.25$ F, and v is the Dirac delta function, we have $V(\omega) = 1$. Hence

$$q(t) = \frac{1}{4\pi} \int_{-\infty}^{\infty} \frac{e^{i\omega t}}{2 - \omega^2 + 2\omega i} \, d\omega$$

$$= -\frac{1}{4\pi} \int_{-\infty}^{\infty} \frac{e^{i\omega t}}{(\omega - \omega_1)(\omega - \omega_2)} \, d\omega \tag{6.18}$$

where $\omega_1 = 1 + i$ and $\omega_2 = -1 + i$. In view of Theorem 5.13,

$$q(t) = -\frac{1}{4\pi} 2\pi i \sum_{i=1}^{2} \text{Res} \left[\frac{e^{i\omega t}}{(\omega - \omega_1)(\omega - \omega_2)}, \omega_i \right]$$

$$= -\frac{i}{2} \left(\frac{e^{i\omega_1 t}}{\omega_1 - \omega_2} + \frac{e^{i\omega_2 t}}{\omega_2 - \omega_1} \right)$$

$$= -\frac{i}{2} \frac{e^{(-1+i)t} - e^{(-1-i)t}}{2}$$

$$= \frac{1}{2} e^{-t} \frac{e^{it} - e^{-it}}{2i}$$

$$= \frac{1}{2} e^{-t} \sin t$$

which is a decaying sinusoidal function.

EXERCISES

6.4.1 Find the charge q flowing in the circuit of Fig. 6.2 when

 a. $v(t) = \delta(t)$, $R = 2\,\Omega$, $L = 1$ H, and $C = 1$ F

 b. $v(t) = \begin{cases} \sin t & \text{for } t \geq 0 \\ 0 & \text{for } t < 0 \end{cases}$

  with $R = 4\,\Omega$, $L = 1$ H, and $C = 1$ F

6.4.2 Compute the Fourier transform of the following functions.

 a. $f(t) = \begin{cases} \cos bt & \text{for } t \geq 0 \\ 0 & \text{for } t < 0 \end{cases}$

 b. $g(t) = u(t) - u(t-1)$ for $t \in \mathbf{R}$ where

$$u(t) = \begin{cases} 1 & \text{for } t \geq 0 \\ 0 & \text{for } t < 0 \end{cases}$$

6.5

LAPLACE TRANSFORMS

Let $f: \mathbf{R} \to \mathbf{C}$ be a (not necessarily continuous) piecewise continuously differentiable function which is not necessarily absolutely integrable on $\mathbf{R}$, but for which there does exist a positive real number σ_0 such that for all $\sigma \geq \sigma_0$, the integral

$$\int_0^\infty e^{-\sigma t} f(t)\, dt$$

converges. An example would be a function of the type $f(t) = \mathcal{O}(e^{\alpha t})$, where by $f(t) = \mathcal{O}(e^{\alpha t})$ we mean that there exists a constant K such that $|f(t)| \leq K e^{\alpha t}$ for all $t \geq 0$ [and where we say that $f(t) = \mathcal{O}(e^{\alpha t})$ is of exponential order $e^{\alpha t}$]; assuming that $\sigma_0 > \alpha$, then if $f(t) = \mathcal{O}(e^{\alpha t})$ and $\sigma \geq \sigma_0$, clearly

$$\int_0^\infty e^{-\sigma t} f(t)\, dt$$

converges. Let u be the "unit step function," that is,

$$u(t) = \begin{cases} 0 & t < 0 \\ 1 & t \geq 0 \end{cases}$$

Set

$$f_1(t) = e^{-\sigma t} f(t) u(t) \qquad \text{for } t \in \mathbf{R} \qquad (6.19)$$

In view of the formula for the inverse Fourier transform, we have

$$f_1(t) = \frac{1}{2\pi} \int_{-\infty}^{\infty} e^{i\omega t} \left[\int_{-\infty}^{\infty} e^{-\sigma t} f(t) e^{-i\omega t} u(t)\, dt \right] d\omega$$

for $t \geq 0$; using Eq. (6.19) and the notation $s = \sigma + i\omega$, we get

$$f(t) = f_1(t)e^{\sigma t}$$

$$= \frac{1}{2\pi} \int_{-\infty}^{\infty} e^{(\sigma + i\omega)t} \left[\int_0^{\infty} f(t)e^{-(\sigma + i\omega)t}\, dt \right] d\omega$$

$$= \frac{1}{2\pi i} \int_{\sigma - i\infty}^{\sigma + i\infty} e^{st} \left[\int_0^{\infty} f(t)e^{-st}\, dt \right] ds$$

The *Laplace transform* of a function is denoted by $F(s) = \mathscr{L}[f](s)$ and defined by

$$F(s) = \mathscr{L}[f](s) = \int_0^{\infty} f(t)e^{-st}\, dt$$

Thus the previous identity implies

$$f(t) = \frac{1}{2\pi i} \int_{\sigma - i\infty}^{\sigma + i\infty} e^{st} \mathscr{L}[f](s)\, ds \qquad\qquad (6.20)$$

The integral in the right-hand side of Eq. (6.20) is called the *inverse Laplace transform* of $\mathscr{L}[f]$.

Let f_1, f_2, $f^{(n)}$, and $\int_{-\infty}^t g(\zeta)\, d\zeta$ be piecewise continuously differentiable functions from $[0,\infty)$ into $\mathbf{C}$ which are of exponential order $e^{\alpha t}$. From the definition of Laplace transform, the following properties hold.

1. *Linearity.* For complex numbers a_1 and a_2,

$$\mathscr{L}[a_1 f_1 + a_2 f_2] = a_1 \mathscr{L}[f_1] + a_2 \mathscr{L}[f_2]$$

2. $\mathscr{L}[f^{(n)}](s) = s^n \mathscr{L}[f] - [s^{n-1}f(0) + s^{n-2}f'(0) + \cdots + f^{(n-1)}(0)]$
$$\text{if Re } s > \alpha.$$

3. $\mathscr{L}\left[\int_{-\infty}^t g(\xi)\, d\xi \right](s) = \dfrac{\mathscr{L}[g(t)](s)}{s} + \dfrac{\int_{-\infty}^0 g(t)\, dt}{s} \qquad \text{if Re } s > \alpha.$

4. If $\mathscr{L}[f_1] = \mathscr{L}[f_2]$, then $f_1(t) = f_2(t)$ at all points at which both f_1 and f_2 are continuous.

In the following example, we indicate the connection between complex variables and Laplace transforms and the usefulness of the latter in solving a mechanical problem.

Example 6.5 Find the angular velocity ω of the rotating mechanical system shown in Fig. 6.3 when a time-varying external torque L is applied. In particular, find ω when $L(t) = u(t)$ for $t \in \mathbf{R}$ and the conditions (6.24) to (6.30) are satisfied.

Solution. Let I be the moment of inertia of the wheel (I is equivalent to mass for linearly moving systems) which is supported by a shaft with torsional compliance K. The brake coefficient of friction D on the side introduces a

Fig. 6.3

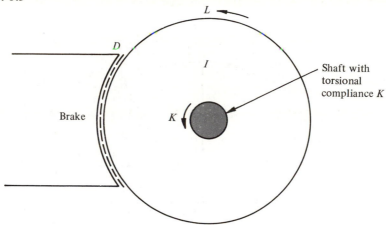

damping into the system. If we denote by L_I the inertial torque, then from mechanics we have

$$L_I + L_D + L_K = L \qquad (6.21)$$

where L is the applied external torque.

Since

$$L_I = I\frac{d\omega}{dt} \qquad L_D = D\omega \qquad \text{and} \qquad L_K = \frac{1}{K}\int_{-\infty}^{t} \omega(\xi)\,d\xi$$

Eq. (6.21) reduces to

$$I\frac{d\omega}{dt} + D\omega + \frac{1}{K}\int_{-\infty}^{t} \omega(\xi)\,d\xi = L \qquad (6.22)$$

Recall that $\omega = d\theta/dt$, where θ is the angular displacement.

By taking the Laplace transform of both sides of (6.22), we obtain (after some manipulations)

$$\left(Is + D + \frac{1}{Ks}\right)\mathscr{L}[\omega](s) = I\omega(0) - \frac{1}{Ks}\int_{-\infty}^{0} \omega(\xi)\,d\xi + \mathscr{L}[L](s) \qquad (6.23)$$

In particular, when $L(t) = u(t)$, and

$$D = 0 \qquad (6.24)$$

$$\int_{-\infty}^{0} \omega(\xi)\,d\xi = \theta_0 \text{ (initial displacement)} = 0 \qquad (6.25)$$

$$\omega(0) = \omega_0 \qquad (6.26)$$

$$I = K = 1 \qquad (6.27)$$

Fig. 6.4

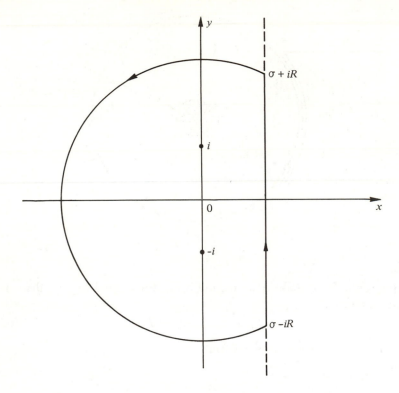

we obtain from Eq. (6.23)

$$\mathscr{L}[\omega](s) = \frac{1}{1 + s^2}$$

Thus

$$\omega(t) = \mathscr{L}^{-1}\left[\frac{1}{1 + s^2}\right] = \frac{1}{2\pi i}\int_{\sigma - i\infty}^{\sigma + i\infty}\frac{e^{st}}{s^2 + 1}\,ds = \sin t$$

The above answer can be found by integrating around the contour shown in Fig. 6.4 (by using the Cauchy residue theorem and then letting $R \to \infty$).

EXERCISES

6.5.1 Compute the Laplace transform of the following functions

    a. $u(t) \sin at$ *b.* $u(t) \cos bt$

    c. $u(t)e^{at}$

6.5.2 Find the charge q flowing in the circuit of Fig. 6.2 when the time-varying voltage source v is given by

$$v(t) = u(t) - u(t - 1) \qquad \text{for } t \in \mathbf{R}$$

and the values of the circuit element are $R = 4\,\Omega$, $L = 2$ H, and $C = 0.25$ F.

Appendices

A.1

INTRODUCTION

In this appendix we state without proof many results from complex function theory, some of which are extremely deep, others trivial, but all of interest. It is our hope that these sections will indicate to the student the flavor of further topics in complex function theory and will stimulate his appetite for further study in this subject. With the aid of some examples and explanations, the student at this point will be of sufficient maturity to fully understand and apply these results. The topics that we have chosen include conformal mappings and Möbius transformations, the Mittag-Leffler and Weierstrass product theorems, doubly periodic functions, sequences of analytic functions, and finally the little and big Picard theorems.

A.2

CONFORMAL MAPPINGS AND MÖBIUS TRANSFORMATIONS

Let $\gamma: [a,b] \to \mathbf{C}$ be a curve. Then if $\gamma'(t_0)$ exists and is nonzero, we call $\gamma'(t_0)$ the *tangent* to γ at $\gamma(t_0)$ (if $t_0 = a$ or $t_0 = b$, we take the corresponding one-sided limit), and we call $\arg \gamma'(t_0)$ the *inclination* of γ at $\gamma(t_0)$. Note that if α is the inclination of γ at $\gamma(t_0)$, then

$$\alpha = \lim_{h \to 0} \arg \frac{\gamma(t + h) - \gamma(t)}{h} \tag{A.1}$$

and

$$e^{i\alpha} = \frac{\gamma'(t_0)}{|\gamma'(t_0)|} \tag{A.2}$$

The following result follows directly from the definitions.

Theorem A.1 Let $D \subset \mathbf{C}$, let $\gamma: [a,b] \to D$ be a curve, and let $f: D \to \mathbf{C}$ be continuous. Also assume $t_0 \in [a,b]$ such that $\gamma'(t_0) \neq 0$ and $f'(\gamma(t_0)) \neq 0$. Finally, let α be the inclination of γ at $\gamma(t_0)$. Then the inclination of $f \circ \gamma$ at $f(\gamma(t_0))$ is $\alpha + \arg f'(\gamma(t_0))$.

Let γ_1 and γ_2 be two curves in $\mathbf{C}$ intersecting at a point z_0, and suppose γ_1 and γ_2 each has a tangent at z_0. We define the *oriented angle* between γ_1 and

γ_2 at z_0 to be the angle through which the tangent of γ_1 at z_0 must be rotated in the counterclockwise direction to reach the tangent of γ_2 at z_0.

Let $D \subset \mathbf{C}$. A continuous function $g\colon D \to \mathbf{C}$ is called *conformal* at z_0 if whenever γ_1 and γ_2 are two curves in D which intersect at z_0 and have tangents at z_0, then the oriented angle between γ_1 and γ_2 at z_0 is equal to the oriented angle between $g \circ \gamma_1$ and $g \circ \gamma_2$ at $g(z_0)$. It is clear by Theorem A.1 that if $f'(z_0) \neq 0$, then f is conformal at z_0.

Recall that an analytic function is also called holomorphic. If $f\colon D \to \mathbf{C}$, then f is *antiholomorphic* if the function h defined by $h(z) = f(\bar{z})$ is holomorphic. A continuous function $g\colon D \to \mathbf{C}$ is called *anticonformal* at z_0 if whenever γ_1 and γ_2 are two curves in D which intersect at z_0 and have tangents at z_0, then the oriented angle between γ_1 and γ_2 at z_0 is equal to the oriented angle between $g \circ \gamma_2$ and $g \circ \gamma_1$ at $g(z_0)$. It is clear by Theorem A.1 that if f is antiholomorphic at z_0 and if $h'(z_0) \neq 0$ [where $h(z) = f(\bar{z})$], then f is anticonformal at z_0.

A continuous function $g\colon D \to \mathbf{C}$ is called *angle-preserving* at z_0 if whenever γ_1 and γ_2 are two curves in D which intersect at z_0 and have tangents at z_0, then the oriented angle between γ_1 and γ_2 at z_0 is equal to either the oriented angle between $g \circ \gamma_1$ and $g \circ \gamma_2$ at $g(z_0)$ or the oriented angle between $g \circ \gamma_2$ and $g \circ \gamma_1$ at z_0.

Recall that $f(z) = u(x,y) + iv(x,y)$ is said to be continuously differentiable as a function of two real variables if u_x, u_y, v_x, and v_y exist and are all continuous. In this case we define the *Jacobian of f at z* to be

$$J(f)(z) = \det \begin{bmatrix} u_x(x,y) & u_y(x,y) \\ v_x(x,y) & v_y(x,y) \end{bmatrix}$$

$$= u_x(x,y)v_y(x,y) - u_y(x,y)v_x(x,y)$$

Note that if f is holomorphic at z, then $J(f)(z) = |f'(z)|^2$, whereas if f is antiholomorphic at z, then $J(f)(z) = -|h'(z)|^2$, where $h(z) = f(\bar{z})$. We are now ready to state a classification theorem for angle-preserving maps.

Theorem A.2 Let D be a domain in $\mathbf{C}$, and let $f\colon D \to \mathbf{C}$ be a continuously differentiable function with nonvanishing Jacobian. Then f is angle-preserving if and only if f is either holomorphic or antiholomorphic throughout D.

Theorem A.3 Let $f\colon D \to \mathbf{C}$ be analytic at z_0 and assume $n \in \mathbf{N}$ such that $f^{(k)}(z_0) = 0$ for $k = 1, \ldots, n-1$, while $f^{(n)}(z_0) \neq 0$. Then if γ_1 and γ_2 are curves in D which intersect at z_0 and have tangents at z_0, then the oriented angle between $f \circ \gamma_1$ and $f \circ \gamma_2$ at $f(z_0)$ is equal to n times the oriented angle between γ_1 and γ_2 at z_0.

As an example, if $f(z) = z^5$ for $z \in \mathbf{C}$, then f preserves the angle between curves passing through $z \neq 0$ and multiplies by 5 the angle between curves passing through 0.

Note that a conformal mapping need not be one-to-one. The following theorem shows, however, that a conformal map is locally one-to-one.

Theorem A.4 Let f be analytic at z_0 and let $f'(z_0) \neq 0$. Then there exists $r > 0$ such that the closed disk $|z - z_0| \leq r$ is contained in dmn f, and moreover such that $f(z) \neq f(z_0)$ when $0 < |z - z_0| \leq r$. Let

$$\delta = \inf_{|z-z_0|=r} |f(z) - f(z_0)|$$

and let $\gamma(t) = z_0 + re^{it}$ for $0 \leq t \leq 2\pi$. Finally, let

$$g: \{w: |w - f(z_0)| < \delta\} \to \mathbf{C}$$

be given by

$$g(w) = \frac{1}{2\pi i} \int_\gamma \frac{zf'(z)}{f(z) - w} \, dz \qquad \text{for } |w - f(z_0)| < \delta$$

Then if $|w - f(z_0)| < \delta$,

$$g(w) = z_0 + \sum_{n=1}^\infty \frac{1}{2\pi n i} \left\{ \int_\gamma \frac{1}{[f(z) - f(z_0)]^n} \, dz \right\} [w - f(z_0)]^n$$

and g is therefore an analytic function. Moreover, $g(f(z)) = z$ whenever $|z - z_0| < r$ and $|f(z) - f(z_0)| < \delta$, and $f(g(w)) = w$ whenever $|w - f(z_0)| < \delta$ and $|g(w) - z_0| < r$. Hence there exists $P > 0$ such that $f|\{z: |z - z_0| < P\}$ is one-to-one.

Corollary A.1 If f is a one-to-one analytic function, then f^{-1} is an analytic function.

Corollary A.2 Let D be a domain and let $f: D \to \mathbf{C}$ be a nonconstant analytic function. Then f is locally an nth-power map; that is, if $z_0 \in D$, then there exist $R > 0$, a positive integer n, and a one-to-one analytic function

$$\phi: \{z: |z - z_0| < R\} \to \mathbf{C}$$

such that if $|z - z_0| < R$, then $f(z) = [\phi(z)]^n$. Hence f is an open map (in other words if O is an open subset of D, then $f[O]$ is an open subset of $\mathbf{C}$).

We say $A \subset \mathbf{C}$ and $B \subset \mathbf{C}$ are *conformally equivalent* if there exists a one-to-one and onto analytic map $f: A \to B$. Note that if one knows all the conformal self-equivalences of A and a particular conformal equivalence of A with B, then by using composition it is clear that one knows all the conformal equivalences of A with B. With this in mind, we shall state the conformal self-equivalences of several spaces.

Theorem A.5 Let $U = \{z: |z| < 1\}$, and let $f: U \to U$ be a conformal self-equivalence. Then there exists $\lambda, z_0 \in \mathbf{C}$ with $|\lambda| = 1$ and $|z_0| < 1$ such that

$$f(z) = \lambda \frac{z - z_0}{\bar{z}_0 z - 1}$$

Moreover, every map of this type is a conformal self-equivalence of U.

Theorem A.6 The conformal self-equivalence of $\mathbf{C}$ are those maps $f(z) = az + b$ for $z \in \mathbf{C}$ where $a, b \in \mathbf{C}$.

We say $f: \tilde{\mathbf{C}} \to \tilde{\mathbf{C}}$ is a conformal self-equivalence of $\tilde{\mathbf{C}}$ if

a. f is a homeomorphism of $\tilde{\mathbf{C}}$ onto $\tilde{\mathbf{C}}$
b. $f|\mathbf{C}$ is meromorphic
c. $f(1/z)$ is meromorphic at 0

Let a, b, c, $d \in \mathbf{C}$ with $ad - bc \neq 0$ and let $f: \tilde{\mathbf{C}} \to \tilde{\mathbf{C}}$ be defined as follows:

$$f(z) = \begin{cases} \dfrac{az + b}{cz + d} & z \in \mathbf{C} - \left\{-\dfrac{d}{c}\right\} \\ \infty & z = -d/c \\ \dfrac{a}{c} & z = \infty \end{cases}$$

Then we call f a *Möbius transformation* (August Ferdinand Möbius, 1790–1868, German), or a *linear fractional transformation*. For simplicity, we often write the above transformation as $w = (az + b)/(cz + d)$.

Theorem A.7 The conformal self-equivalences of $\tilde{\mathbf{C}}$ are precisely the Möbius transformations. Moreover if p_1, p_2, and p_3 are distinct points in $\tilde{\mathbf{C}}$ and if q_1, q_2, and q_3 are distinct points in $\tilde{\mathbf{C}}$, then there exists a unique Möbius transformation $f: \tilde{\mathbf{C}} \to \tilde{\mathbf{C}}$ such that $f(p_j) = q_j$ for $j = 1, 2, 3$.

Theorem A.8 Let a, b, c, $d \in \mathbf{R}$ such that $ad - bc > 0$. Then if $f(z) = (az + b)/(cz + d)$ for $z \in \{w: \text{Im } w > 0\}$, then f is a conformal equivalence of $\{w: \text{Im } w > 0\}$ with itself, and moreover, every conformal self-equivalence of $\{w: \text{Im } w > 0\}$ arises this way.

AUGUST FERDINAND MÖBIUS

August Ferdinand Möbius was born at Schulpforte, Prussia, on November 17, 1790. He studied at the universities of Leipzig, Göttingen, and Halle. In 1815 he was appointed professor of astronomy and director of the observatory in Leipzig.

Möbius wrote several books on astronomy as well as a number of works on various topics in analytical geometry. He was a pioneer in topology; in a memoir discovered after his death, the properties of the now-famous Möbius strip are described. He died in Leipzig on September 26, 1868.

Thus we see that every conformal equivalence of the open unit disk, the upper half-plane, and the complex plane extends to one of the extended complex plane.

Next, we shall state one of the most beautiful results in complex functions, together with two results which follow from it.

Theorem A.9. *Riemann Mapping Theorem* Let A be a nonempty simply connected domain in U. Then A and U are conformally equivalent.

Corollary A.3 Let A be a nonempty simply connected domain in $\mathbf{C}$, and let $\mathbf{C} - A \neq \emptyset$. Then A is conformally equivalent to U.

Corollary A.4 Let A and B be simply connected nonempty subsets of $\mathbf{C}$. Then A and B are homeomorphic.

Note that Corollary A.4 states that U and $\mathbf{C}$ are homeomorphic. However, Liouville's theorem clearly implies that they are not conformally equivalent.

Finally we shall state some elementary properties of Möbius transformations and present some examples which can be easily verified.

By a *generalized circle* we mean either an ordinary circle (that is,

$$\{z \in \mathbf{C}: |z - z_0| = R\}$$

for some $z_0 \in \mathbf{C}$ and some positive real number R) or a straight line. Then we have the following theorem.

Theorem A.10 Möbius transformations map generalized circles onto generalized circles.

Example A.1 The Möbius transformation

$$w = \frac{z - 1}{z - i}$$

maps the circle $|z| = 2$ onto the circle $|w - (2 + i)| = \sqrt{5}$, and the circle $|z| = 1$ onto the generalized circle (straight line) $w - i\bar{w} = 0$.

Proof. Solving the equation $w = (z - 1)/(z - i)$ with respect to z, we obtain $z = (iw - 1)/(w - 1)$. Thus the circle $|z| = 2$ is mapped onto

$$\left| \frac{iw - 1}{w - 1} \right| = 2$$

that is, the circle

$$\left| w - \frac{4 + i}{3} \right| = \frac{2\sqrt{2}}{3}$$

Also the circle $|z| = 1$ is mapped onto

$$\left| \frac{iw - 1}{w - 1} \right| = 1$$

which is the straight line $w - i\bar{w} = 0$.

The *cross ratio* of four distinct numbers z_1, z_2, z_3, and z_4 is denoted by (z_1, z_2, z_3, z_4) and is defined by

$$(z_1, z_2, z_3, z_4) = \frac{(z_1 - z_2)(z_3 - z_4)}{(z_1 - z_4)(z_3 - z_2)}$$

If one of the four distinct complex numbers, say z_1, is the point at infinity, their cross ratio is defined as the limit of (z, z_2, z_3, z_4) as $z \to \infty$. With this definition in mind and if

$$w_i = \frac{az_i + b}{cz_i + d}$$

for $i = 1, 2, 3, 4$, it is easily seen that (w_1, w_2, w_3, w_4) is defined and

$$(w_1, w_2, w_3, w_4) = (z_1, z_2, z_3, z_4)$$

Thus we have the following theorem.

Theorem A.11 The cross ratio of four distinct points is preserved under a Möbius transformation.

The above theorem can be used to find the Möbius transformation that maps three given points into the three preassigned points.

Example A.2 Find the Möbius transformation mapping the points 1, -1, and i into 0, ∞, and 1.

Solution. If the desired transformation takes the point z into the point w, then by Theorem A.11 we have

$$(0, \infty, 1, w) = (1, -1, i, z)$$

Therefore

$$\frac{1 - w}{0 - w} = \frac{2(i - z)}{(1 - z)(i + 1)}$$

and so

$$w = \frac{z - 1}{i(z + 1)}$$

Let $|z - z_0| = R$ be a circle in $\mathbf{C}$ with center z_0 and radius R. Two points z_1 and z_2 collinear with z_0 are called *inverse points* with respect to the circle $|z - z_0| = R$ if

$$|z_1 - z_0|\, |z_2 - z_0| = R^2$$

whereas two points z_1 and z_2 are called inverse points with respect to a straight line if z_1 and z_2 are symmetric with respect to the straight line.

Theorem A.12 The symmetric point z_2 of a point z_1 with respect to the circle $|z - z_0| = R$ is given by

$$z_2 = z_0 + \frac{R^2}{\bar{z}_1 - \bar{z}_0}$$

Theorem A.13 A Möbius transformation $w = (az + b)/(cz + d)$ transforms two inverse points into two points which are also inverse with respect to the transformed circle.

Theorem A.14 A one-to-one analytic function that maps the interior of a circle onto the interior of another circle is a Möbius transformation.

Using the above three theorems one can establish the following facts.

1. The transformation

$$w = \lambda \frac{z - z_0}{z - \bar{z}_0} \qquad |\lambda| = 1$$

is the most general Möbius transformation mapping the upper half-plane into the unit circle $|z| = 1$ in such a way that the point z_0 of the upper half-plane is mapped into the center of the circle.

2. The transformation

$$w = \lambda \frac{z - z_0}{1 - \bar{z}_0 z} \qquad |\lambda| = 1$$

is the most general analytic function that maps the unit circle $|z| < 1$ onto itself in such a way that the point $z_0 \in \{z \in \mathbf{C}: |z| < 1\}$ is mapped into the center of the unit circle.

A.3
THE MITTAG-LEFFLER AND
WEIERSTRASS PRODUCT THEOREMS

In this section we examine the problem of constructing functions with preassigned zeros or poles. The first theorem is due to Gösta Mittag-Leffler (1846–1927, German).

Theorem A.15 Let A be an open subset of $\mathbf{C}$ and let $z_1, z_2, \ldots$ be a sequence of distinct points in A having no limit points in A. For each $n \in \mathbf{N}$, let ψ_n be a polynomial such that $\psi_n(0) = 0$. Then there exists an analytic function $f: A - \{z_n: n = 1, 2, \ldots\} \to \mathbf{C}$ such that f is meromorphic on A, and such that if $n \in \mathbf{N}$, then the principal part of f at z_n is given by

$$P_{z_n}(z) = \psi_n\left(\frac{1}{z - z_n}\right) \qquad \text{for } z \in \mathbf{C} - \{z_n\}$$

Theorem A.16 Let A be an open subset of $\mathbf{C}$, and let f be meromorphic on A such that $A - \operatorname{dmn} f = \{z_1, z_2, \ldots\}$, where $z_1, z_2, \ldots$ is a sequence of distinct points in A having no limit points in A. Then there exist analytic functions $g: A \to \mathbf{C}$ and $W_n: A - \{z_n\} \to \mathbf{C}$ for $n = 1, 2, \ldots$ such that for each n, W_n is meromorphic at z_n and has the same principal part at z_n as f, such that $\sum_{n=1}^{\infty} W_n(z)$ is absolutely convergent if $z \in A - \{z_1, z_2, \ldots\}$, and such that if $z \in A - \{z_1, z_2, \ldots\}$, then $f(z) = g(z) + \sum_{n=1}^{\infty} W_n(z)$.

The following special case of Theorem A.16 is often useful in computing a meromorphic function f on $\mathbf{C}$ having a preassigned set of simple poles and residues.

Theorem A.17 Let f be a meromorphic function on $\mathbf{C}$ whose nonzero poles are the distinct points $z_1, z_2, \ldots$ with $\operatorname{Res}(f, z_n) = a_n$ for $n = 1, 2, \ldots$. If 0 is a pole of f, let $\operatorname{Res}(f, 0) = a_0$, while if 0 is not a pole of f, let $a_0 = 0$. Assume that for some nonnegative integer k, the series

$$\sum_{n=1}^{\infty} \frac{|a_n|}{|z_n|^{k+1}}$$

converges. Finally, assume that all poles of f are simple. Then there exists an entire function $g: \mathbf{C} \to \mathbf{C}$ such that if $z \in \mathbf{C} - \{0, z_1, z_2, \ldots\}$, then

$$f(z) = g(z) + \frac{a_0}{z} + \sum_{n=1}^{\infty} \left(\frac{z}{z_n}\right)^k \frac{a_n}{z - z_n}$$

Example A.3 Show that there exists an entire function $g: \mathbf{C} \to \mathbf{C}$ such that if $z \in \mathbf{C} - \mathbf{Z}$, then

$$\frac{\pi}{\sin \pi z} = g(z) + \frac{1}{z} + \sum_{-\infty < n < +\infty} (-1)^n \left(\frac{1}{z - n} + \frac{1}{n} \right)$$

Proof. Here $f(z) = \pi/(\sin \pi z)$ for $z \in \mathbf{C} - \mathbf{Z}$, and therefore the poles of f are the points $0, \pm 1, \pm 2, \ldots$ with Res $(f, n) = (-1)^n = a_n$ for $n = 0, \pm 1, \pm 2, \ldots$. The result follows from Theorem A.17 and the observation that in this example we can take $k = 1$. It can be shown that the function g of Example A.3 is identically equal to zero.

The next theorem is called the *Weierstrass product theorem*.

Theorem A.18 Let A be an open subset of $\mathbf{C}$, and let $z_1, z_2, \ldots$ be a sequence of distinct points in A having no limit points in A. For each $n \in \mathbf{N}$, let v_n be a positive integer. Then there exists an analytic function $f: A \to \mathbf{C}$ such that $\{z \in A : f(z) = 0\} = \{z_1, z_2, \ldots\}$ and such that $\mathcal{O}(f, z_n) = v_n$ for $n = 1, 2, \ldots$.

Before we give the next corollary, it will be necessary to define what we mean by $\prod_{n=1}^{\infty} h_n$. Let $h_1, h_2, \ldots$ be a sequence of complex numbers. Then the symbol $\prod_{n=1}^{\infty} h_n$ is defined in the following two cases.

1. Suppose $h_k = 0$ for some $k \in \mathbf{N}$. Then we set $\prod_{n=1}^{\infty} h_n = 0$.
2. Suppose $h_k \neq 0$ for all $k \in \mathbf{N}$. Further suppose there exists a sequence $l_1, l_2, \ldots$ of complex numbers with $\sum_{n=1}^{\infty} l_n$ absolutely convergent such that $e^{l_n} = h_n$ for all $n \in \mathbf{N}$. Then we define $\prod_{n=1}^{\infty} h_n = e^l$, where $\sum_{n=1}^{\infty} l_n = l$.

Theorem A.19 Let D be a domain and let $f: D \to \mathbf{C}$ be a nonconstant analytic function such that $\{z : f(z) = 0\} = \{z_1, z_2, \ldots\}$, where $z_1, z_2, \ldots$ is a sequence of distinct points in D having no limit points in D. Then there exist analytic functions $g: D \to \mathbf{C}$ and $w_n: D \to \mathbf{C}$ for $n = 1, 2, \ldots$ such that $g(z) \neq 0$ for all $z \in D$, $w_n(z) \neq 0$ if $z \neq z_n$, $\mathcal{O}(w_n, z_n) = \mathcal{O}(f, z_n)$, and finally such that $f(z) = g(z) \prod_{n=1}^{\infty} w_n(z)$ if $z \in D$.

Corollary A.5 Let f be meromorphic on A. Then there exist analytic functions $g: A \to \mathbf{C}$ and $h: A \to \mathbf{C}$ such that $f = g/h$.

Corollary A.6 Let D be a domain. Then the ring of all analytic functions on D is an integral domain whose field of quotients is isomorphic to the field of those functions meromorphic on D with no removable singularities in D.

The following theorem is often useful in computing an analytic function f on a domain D having preassigned zeros and multiplicities.

Theorem A.20. *The Logarithmic Derivative Theorem* Let D be a domain in $\mathbf{C}$, let $H = \{f : f$ is analytic in $D\}$, and let $L = \{g : g$ is meromorphic on A, all poles of g in A are simple, all residues of g are positive integers, and $1/(2\pi i) \int_\gamma g(z)\, dz \in \mathbf{Z}$ whenever γ is a closed curve in dmn $g\}$. Then the following hold.

a. If $f \in H$ and $g = f'/f$, then $g \in L$ and

$$\exp\left[\int_\gamma g(z)\, dz\right] = \frac{f(\gamma(b))}{f(\gamma(a))}$$

for every curve $\gamma : [a,b] \to$ dmn f.
b. If $g \in L$, there exists $f \in H$ such that $f'/f = g$.
c. If f_1 and $f_2 \in H$, then $f_1'/f_1 = f_2'/f_2$ if and only if there exists $\lambda \in \mathbf{C} - \{0\}$ such that $f_1 = \lambda f_2$.
d. If f_1 and $f_2 \in H$, then

$$\frac{(f_1 f_2)'}{f_1 f_2} = \frac{f_1'}{f_1} + \frac{f_2'}{f_2}$$

e. If $f_1, f_2, \ldots \in H$, and if $\zeta \in D$ such that $f_n(\zeta) = 1$ for $n = 1, 2, \ldots$, and if there exists g meromorphic on A such that $\zeta \in$ dmn g, and such that for every compact set $B \subset$ dmn g,

$$\lim_{n\to\infty} \sup_{z\in B} \left| \frac{f_n'(z)}{f_n(z)} - g(z) \right| = 0$$

then there exists $f \in H$ such that $f'/f = g$ and such that for every compact set $B \subset A$,

$$\lim_{n\to\infty} \sup_{z\in B} |f_n(z) - f(z)| = 0$$

Thus we see by part (a) above that if f is analytic on D with zeros $z_1, z_2, \ldots$ and corresponding multiplicities $\mathcal{O}(f, z_n) = a_n$ for $n = 1, 2, \ldots$, then $g = f'/f$ is analytic on $D - \{z_1, z_2, \ldots\}$ and has simple poles at $z_1, z_2, \ldots$ with residues Res $(g, z_n) = a_n$ for $n = 1, 2, \ldots$. If $z_0 \in D - \{z_1, z_2, \ldots\}$, then for all $z \in D - \{z_1, z_2, \ldots\}$, while the integral

$$\int_{z_0}^{z} g(\zeta)\, d\zeta$$

is not well defined, we can see that

$$\exp\left[\int_{z_0}^{z} g(\zeta)\, d\zeta\right]$$

is well defined, and in fact, $f(z) = f(z_0) \exp\left[\int_{z_0}^{z} g(\zeta)\, d\zeta\right]$.

Example A.4 Show that

$$\sin \pi z = \pi z \prod_{n=1}^{\infty} \left(1 - \frac{z^2}{n^2}\right)$$

Proof. The function

$$f(z) = \begin{cases} \dfrac{\sin \pi z}{\pi z} & z \neq 0 \\ 1 & z = 0 \end{cases}$$

is an entire function with simple zeros at the points $\pm 1, \pm 2, \ldots$. Clearly,

$$g(z) = \frac{f'(z)}{f(z)} = -\frac{1}{z} + \pi \cot \pi z$$

is a meromorphic function on **C** with simple poles at $\pm 1, \pm 2, \ldots$ and Res $(g,n) = 1$. Here, as in Example A.3 we can take $k = 1$, and from Theorem A.17 we can obtain

$$\frac{f'(z)}{f(z)} = g(z) = \sum_{\substack{-\infty < n < +\infty \\ n \neq 0}} \left(\frac{1}{z - n} + \frac{1}{n} \right)$$

By integrating the last equality from 0 to z, the result follows.

A.4
DOUBLY PERIODIC FUNCTIONS

In Sec. 3.4 we defined the following concepts: periodic function, primitive period, and simply periodic function. Let f be a meromorphic periodic function on **C** with a primitive period ω_1. Then the set $S = \{n\omega_1 : n = \pm 1, \pm 2, \ldots \}$ lies on the line L passing through 0 and ω_1, and any point of S is a period of f. Assume now that ω_2 is a primitive period of f which does not lie on L. It is easily seen that any point of the set

$$T = \{n\omega_1 + m\omega_2 : n, m \in \mathbf{Z} \quad \text{and} \quad |n| + |m| \neq 0\}$$

is a period of f. Conversely, one can show that the only periods of f are the points of the set T. A meromorphic function $f: D \to \mathbf{C}$ which has two primitive periods ω_1 and ω_2 such that the points 0, ω_1, and ω_2 are not collinear is called a *doubly periodic* function or an *elliptic* function. Clearly a doubly periodic function is well determined by its values on a *period parallelogram*, that is, any parallelogram with vertices z_0, $z_0 + \omega_1$, $z_0 + \omega_1 + \omega_2$, $z_0 + \omega_2$, where $z_0 \in \mathbf{C}$. If P is a period parallelogram of the elliptic function f, then the sum of the orders of the poles of f in P is called the *order* of P (which by Theorem 5.3 is well defined). Let us state some facts about elliptic functions.

1. If the order of the elliptic function f is zero, then f is a constant function.
2. If f_1 and f_2 are elliptic functions, then so are the functions
   a. $c_1 f_1 + c_2 f_2 + c_3 f_1' + c_4 f_2'$, where $c_1, c_2, c_3, c_4 \in \mathbf{C}$
   b. f_1/f_2, if $f_2(z) \neq 0$ for all $z \in \text{dmn } f_2$
   c. $f_1 f_2$
3. The sum of the residues of an elliptic function in any period parallelogram P is zero. Consequently, there do not exist elliptic functions of order one.

An important example of an elliptic function is the Weierstrass $\mathscr{P}$ function. Let ω_1 and ω_2 be two complex numbers so that 0, ω_1, and ω_2 are not collinear. Consider the set $S = \{n\omega_1 + m\omega_2 : n, m \in \mathbf{Z}\}$. The Weierstrass $\mathscr{P}$ function is a meromorphic function with double poles at each point of S, analytic in $\mathbf{C} - S$, and with the principal part of $\mathscr{P}$ at $n\omega_1 + m\omega_2$ being

$$\frac{1}{(z - n\omega_1 - m\omega_2)^2}$$

It can be shown that $\mathscr{P}$ is an elliptic function given by the absolutely convergent series

$$\mathscr{P}(z : \omega_1, \omega_2) = \frac{1}{z^2} + \sum_{\substack{n, m \\ |n| + |m| \neq 0}} \left[\frac{1}{(z - n\omega_1 - m\omega_2)^2} - \frac{1}{(n\omega_1 + m\omega_2)^2} \right]$$

The following result is of great interest in the theory of elliptic functions.

Theorem A.21 Assume that $\omega_1/\omega_2 \notin \mathbf{R}$. A function f is elliptic with periods ω_1 and ω_2 if and only if f can be represented rationally in terms of $\mathscr{P}(z : \omega_1, \omega_2)$ and $\mathscr{P}'(z : \omega_1, \omega_2)$.

EXERCISES

A.3.1 Show that the derivative of the Weierstrass $\mathscr{P}$ function is doubly periodic.

A.3.2 Show that the Weierstrass $\mathscr{P}$ function is doubly periodic.

A.5
SEQUENCES OF ANALYTIC FUNCTIONS

Throughout this section, we assume that D is a domain and that $f_1, f_2, \ldots$ is a sequence of functions analytic on D.

Theorem A.22 Suppose that for all compact subsets $K \subset D$,

$$\lim_{m, n \to \infty} \sup_{z \in K} |f_n(z) - f_m(z)| = 0$$

Then there exists an analytic function $g : D \to \mathbf{C}$ such that if K is a compact subset of D, then

$$\lim_{n \to \infty} \sup_{z \in K} |g(z) - f_n(z)| = 0$$

and such that if k is a positive integer, then

$$\lim_{n \to \infty} \sup_{z \in K} |g^{(k)}(z) - f_n^{(k)}(z)| = 0$$

Theorem A.23 Suppose that

$$\lim_{m,n \to \infty} \sup_{z \in D} |f_m(z) - f_n(z)| = 0$$

Then the function g from Theorem A.22 actually satisfies

$$\lim_{n \to \infty} \sup_{z \in D} |g(z) - f_n(z)| = 0$$

The fact that Theorem A.23 does not yield a sharper result about derivatives than Theorem A.22 is shown in the following example.

Example A.5 Let

$$f_n(z) = \sum_{k=1}^{n} \frac{1}{k^2} z^k \qquad \text{for } |z| < 1$$

Then

$$g(z) = \sum_{n=1}^{\infty} \frac{1}{n^2} z^n$$

converges uniformly on the disk $|z| < 1$, while

$$g'(z) = \sum_{n=1}^{\infty} \frac{1}{n} z^{n-1}$$

does not. The following theorem is due to Montel.

Theorem A.24 Suppose there exists $M \geq 0$ such that $|f_n(z)| \leq M$ for all $z \in D$ and $n \in \mathbf{N}$. Then there exists a subsequence $f_{n_1}, f_{n_2}, \ldots$ and an analytic function $g: D \to \mathbf{C}$ such that for all compact subsets K of D,

$$\lim_{k \to \infty} \sup_{z \in K} |f_{n_k}(z) - g(z)| = 0$$

Theorem A.25 Suppose there exists $M \geq 0$ such that $|f_n(z)| \leq M$ for all $z \in D$ and $n \in \mathbf{N}$. Let

$$S = \{z \in D: \text{the sequence } f_1(z), f_2(z), \ldots \text{ is convergent}\}$$

Then if S is not discrete, there exists an analytic function $g: D \to \mathbf{C}$ such that for all compact subsets K of D,

$$\lim_{n \to \infty} \sup_{z \in K} |f_n(z) - g(z)| = 0$$

The following example shows the importance of the existence of the number M in Theorems A.24 and A.25.

Example A.6 Let $f_n(z) = 2^n \sin(2^n \pi z)$ for $z \in \mathbf{C}$ and $n \in \mathbf{N}$. Then if $m = 0, 1, 2, \ldots, n - 1$, we have

$$f_n\left(\frac{1}{2^m}\right) = 2^n \sin(2^{n-m}\pi) = 0$$

and so if $m \in \mathbf{N}$, then $f_n(1/2^m) \to 0$ as $n \to \infty$. Hence S contains the cluster point 0 and so is not discrete. However,

$$f_n\left(\frac{1}{2^{n+1}}\right) = 2^n \sin\frac{\pi}{2} = 2^n \qquad \text{for } n = 1, 2, \ldots$$

and so there does exist an analytic function $g: D \to \mathbf{C}$ satisfying the conclusions of either Theorem A.24 or A.25. Consternation is avoided by noticing the nonexistence of the constant M required in Theorems A.24 and A.25.

We conclude this section with a theorem about sequences of one-to-one analytic functions.

Theorem A.26 Suppose $f_1, f_2, \ldots$ is a sequence of one-to-one analytic functions, and that there exists a nonconstant function $g: D \to \mathbf{C}$ such that for all compact subsets K of D,

$$\lim_{n \to \infty} \sup_{z \in K} |f_n(z) - g(z)| = 0$$

Then g is a one-to-one analytic function. Furthermore if K is a compact subset of im g, then there exists $N \in \mathbf{N}$ such that if $n \geq N$, $K \subset \text{im } f_n$, and moreover,

$$\lim_{n \to \infty} \sup_{z \in K} |f_n^{-1}(z) - g^{-1}(z)| = 0$$

The necessity of the requirement that g be nonconstant is illustrated by the following example.

Example A.7 Let $f_n(z) = z/n$ for all $z \in \mathbf{C}$ and $n \in \mathbf{N}$. Then if $g(z) = 0$ for all $z \in \mathbf{C}$, we see that for all compact subsets K of $\mathbf{C}$,

$$\lim_{n \to \infty} \sup_{z \in K} |f_n(z) - g(z)| = 0$$

while g is not one-to-one.

A.6
THE LITTLE AND BIG PICARD THEOREMS

In this section we state several deep results, two of which are due to Émile Picard (1856–1941, French). The first theorem states that an entire function misses at most one point.

Theorem A.27. *The Little Picard Theorem* Let $f: \mathbf{C} \to \mathbf{C}$ be a nonconstant entire function. Then $\mathbf{C} - \operatorname{im} f$ has at most one element.

The entire function $f(z) = e^z$ shows that an entire function may indeed miss a point (in this case zero), while Theorem A.27 implies that if $w \in \mathbf{C} - \{0\}$, then there does indeed exist $z \in \mathbf{C}$ such that $e^z = w$. Looking at the entire function $f(z) = e^{e^z}$, a student might find it plausible that f misses two points, namely 0 and also 1, as the exponent $e^z \neq 0$ for all $z \in \mathbf{C}$. We leave it to the student to resolve this apparent contradiction.

The next theorem examines the behavior of an analytic function near a pole.

Theorem A.28 Let $R > 0$ and let $f: \{z: 0 < |z| < R\} \to \mathbf{C}$ be an analytic function. Then the following two conditions are equivalent.

a. f is meromorphic at 0.
b. For each $\zeta \in \mathbf{C}$ with $|\zeta| = 1$, there exists $\delta > 0$ and two distinct points α and $\beta \in \mathbf{C}$ such that

$$f\left[\left\{z: 0 < |z| < \delta \quad \text{and} \quad \left|\frac{z}{|z|} - \zeta\right| < \delta\right\}\right] \subset (\mathbf{C} - \{\alpha, \beta\})$$

The next corollary which examines the behavior of an analytic function near an essential singularity is an immediate consequence of Theorem A.28.

Corollary A.7 Let $R > 0$ and let $f: \{z: 0 < |z| < R\} \to \mathbf{C}$ be an analytic function. Then the following two conditions are equivalent.

a. 0 is an essential singularity of f.
b. There exists $\zeta \in \mathbf{C}$ with $|\zeta| = 1$ and $\alpha \in \mathbf{C}$ such that if $0 < \delta \leq R$,

$$\mathbf{C} - \{\alpha\} \subset f\left[\left\{z: 0 < |z| < \delta \quad \text{and} \quad \left|\frac{z}{|z|} - \zeta\right| < \delta\right\}\right]$$

CHARLES ÉMILE PICARD

Émile Picard was born in Paris, France, on July 24, 1856. As a teacher in several French universities, he contributed significantly to complex function theory and the theory of differential equations. In 1879 he proved the theorem known by his name; it prompted many important investigations by other mathematicians. In 1899 Picard visited the United States as visiting lecturer at Clark University, Worcester, Massachusetts. He died in Paris on December 11, 1941.

Fig. A.1

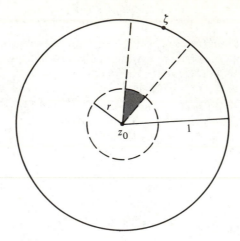

Corollary A.7 merely states the fact that the image under f of the shaded sector in Fig. A.1 misses at most one point. The line $\{z_0 + ty: t > 0\}$ is called a *direction of Julia* (Gaston Julia, 1893–present, French).

In view of Corollary A.7, the reader can easily prove the following result.

Theorem A.29. *The Big Picard Theorem*   Let $z_0 \in \mathbf{C}$, let O be an open set in $\mathbf{C}$, let $z_0 \in O$, and let $f: O - \{z_0\} \to \mathbf{C}$ be analytic and have z_0 as an essential singularity. Then there exists $\alpha \in \mathbf{C}$ such that if $w \in \mathbf{C} - \{\alpha\}$, then $f(z) = w$ for infinitely many $z \in O - \{z_0\}$.

Hints and Answers to
Selected Exercises

Chapter 1

1.2.1 $(1,2)$ is a root since $(1,2)^2 - (2,0)(1,2) + (5,0) = (-3,4) - (2,4) + (5,0) = (0,0)$.

1.2.4 If $z_1 = x_1 + iy_1$ and $z_2 = x_2 + iy_2$, then $z_1 + z_2 = (x_1 + x_2) + i(y_1 + y_2)$ from which (a) and (b) follow, while

$$z_1 z_2 = (x_1 x_2 - y_1 y_2) + i(x_1 y_2 + x_2 y_1)$$

from which (c) and (d) follow.

1.2.5 (a) $-1 + 2i$

1.2.6 $(z^2 - 2z + 2)(z^2 + 1)$

1.2.7 $z^4 = x^4 + 4x^3 iy + 6x^2(iy)^2 + 4x(iy)^3 + (iy)^4$
$= (x^4 - 6x^2 y^2 + y^4) + i(4x^3 y - 4xy^3)$.
Solve the equations Im $z^4 = 0$ and Re $z^4 = 0$.

1.3.2 If $z = x + iy$, then $z^2 = (\bar{z})^2$ if and only if $x^2 - y^2 + 2xyi = x^2 - y^2 - 2xyi$ if and only if $xy = 0$.

1.3.3 $\overline{1 - i} = 1 + i$ is also a root.

1.3.7 Squaring both sides of $|(z - a)/(1 - \bar{a}z)| \le 1$ and simplifying yields $(1 - |z|^2)(1 - |a|^2) \ge 0$.

1.3.8 $|\lambda_1 z_1 + \lambda_2 z_2 + \cdots + \lambda_n z_n| \le |\lambda_1|\,|z_1| + |\lambda_2|\,|z_2| + \cdots + |\lambda_n|\,|z_n| < \lambda_1 + \lambda_2 + \cdots + \lambda_n = 1$

1.3.9 Find λ such that $F'(\lambda) = 0$. [Observe that $F''(\lambda) < 0$.]

1.3.11 When for some $\lambda, \mu \in \mathbf{C}$ we have $\lambda a_j = \mu b_j$ for $j = 1, 2, \ldots, n$.

1.4.2 $2^3\sqrt{3} + 2^3 i; \pi/6$

1.4.4 (b) $|z - (-2 + 5i)| = 3$

1.4.6 The triangles with vertices z_1, z_2, z_3 and z_2, z_3, z_1 are similar. Expand the determinant in Example 1.10 after you replace $\zeta_1, \zeta_2, \zeta_3$ by z_2, z_3, z_1, respectively.

1.4.8 Translate the relations into geometric language and recall your high school geometry.

1.4.9 If z_1, z_2, z_3 are the vertices of the triangle and

$$\zeta_1 = \frac{z_2 - z_1}{z_3 - z_1} \qquad \zeta_2 = \frac{z_3 - z_2}{z_1 - z_2} \qquad \zeta_3 = \frac{z_1 - z_3}{z_2 - z_3}$$

then the angle at z_j is $\pm \operatorname{Arg} \zeta_j$ for $j = 1, 2, 3$. Since $1/\zeta_1 + \zeta_2 = 1$, it follows that Im ζ_1 and Im ζ_2 have the same sign. Thus the sum of the angles is $\pm$ Arg $(\zeta_1\zeta_2\zeta_3)$ mod 2π. The result follows from the facts that $\zeta_1\zeta_2\zeta_3 = -1$ and that each angle is between 0 and π.

1.4.11 To show (c), use the triangle inequality for complex numbers.

1.5.1 The four fourth roots of i are $\frac{1}{2}\left(\pm\sqrt{2+\sqrt{2}} \pm i\sqrt{2-\sqrt{2}}\right)$.

1.5.3 We must show that each root of $(z+1)^5 + z^5 = 0$ has real part equal to $-\frac{1}{2}$. To solve this equation, set $\zeta = (z+1)/z$ and note that $\zeta^5 = -1$.

1.5.4 $|z|^n = |z^n| = |\bar{z}| = |z|$. Thus either $z = 0$ or $|z| = 1$. For $z \neq 0$ the equation is equivalent to $z^{n+1} = \bar{z}z = 1$.

1.6.2 If $p = -3$, then $t_1 = t_2 = -1$ and so by Cardan's formulas $w_1 = -2$, and $w_2 = w_3 = 1$. Let $p < -3$. Then t_1 and t_2 are complex conjugates, and if $\sqrt[3]{t_1} = a + ib$, then $\sqrt[3]{t_2} = a - ib$ (Why?). Thus $w_1 = 2a$, $w_2 = \omega(a+bi) + \omega^2(a-bi) = -a - b\sqrt{3}$, and $w_3 = -a + b\sqrt{3}$.

1.6.3 After setting $z = \tan\theta$, the equation becomes $z^3 + 3z - 3t = 0$. Thus $t_1 = 3t/2 + \sqrt{9t^2/4 + 1}$ and $t_2 = 3t/2 - \sqrt{9t^2/4 + 1}$ are real. If $\sqrt[3]{t_1}$ and $\sqrt[3]{t_2}$ denote real roots of t_1 and t_2, respectively, then $w_1 = \sqrt[3]{t_1} + \sqrt[3]{t_2}$ is real while $w_2 = \omega\sqrt[3]{t_1} + \omega^2\sqrt[3]{t_2} = -(\sqrt[3]{t_1} + \sqrt[3]{t_2})/2 + i\sqrt{3}(\sqrt[3]{t_1} - \sqrt[3]{t_2})/2 = \bar{w}_3 \notin \mathbf{R}$.

1.6.4 (a) If z_1 and z_2 are the roots, then $z_1 + z_2 = -p$ while $z_1z_2 = q$. Now if $p^2 - 4q < 0$, then the roots are complex conjugates, and moreover, $\operatorname{Re} z_1 = \operatorname{Re} z_2 = (z_1 + z_2)/2 = -p/2 < 0$ if and only if $p > 0$. (Why is $q > 0$ here?) If $p^2 - 4q \geq 0$, then both roots are real. They have the same sign if and only if $q > 0$, and they are both negative if and only if $-p < 0$.

(b) Set $z^3 + pz^2 + qz + r = (z + a)(z^2 + bz + c)$ where $p = a + b$, $q = ab + c$, and $r = ac$; use (a).

(c) Set $z^4 + pz^3 + qz^2 + rz + s = (z^2 + az + b)(z^2 + cz + d)$. Find a, b, c, and d and use (a).

Chapter 2

2.2.1 Assume G is open. If G is empty, then G is the union of the empty class of ε neighborhoods. If $G \neq \varnothing$, then every point in G is an interior point, and so every point z in G is in the center of some ε neighborhood $N_{\varepsilon(z)}(z)$ contained in G with radius $\varepsilon(z)$. Clearly $G = \bigcup_{z \in G} G_{\varepsilon(z)}(z)$. Conversely, if $G = \bigcup_{j \in J} N_{\varepsilon_j}(z_j)$, then every point $z \in G$ belongs to some $N_{\varepsilon_j}(z_j)$, and since an ε neighborhood is an open set, z is an interior point of G.

2.2.2 It is clear from Exercise 2.2.1 that any union of open sets is open. Now let $G_1, G_2, \ldots, G_k$ be a finite number of open sets and let $z \in \bigcap_{j=1}^{k} G_j$. Then $z \in G_j$ for $j = 1, 2, \ldots, k$. Since G_j is open, there exists a radius ε_j such that $N_{\varepsilon_j}(z) \subset G_j$ for each $j = 1, 2, \ldots, k$. Set $\varepsilon = \min(\varepsilon_1, \varepsilon_2, \ldots, \varepsilon_k)$. Then $N_\varepsilon(z) \subset \bigcap_{j=1}^{k} G_j$, and so z is an interior point of $\bigcap_{j=1}^{k} G_j$.

2.2.4 Use Exercise 2.2.1 and De Morgan's laws (the complement of a union of sets is the intersection of the complements of the sets, and the complement of the intersection of sets is the union of the complements of the sets).

2.2.7 Use Exercises 2.2.4 and 2.2.5.

2.2.8 Show that their complements are open.

2.2.9 The sets in question are clearly bounded, and by Exercise 2.2.4, they are closed.

2.2.10 It is closed (since its complement is open) and it is clearly bounded.

2.3.1 (*a*) Does not exist. (As $z \to 0$ through real values, the limit is 1, while as $z \to 0$ through imaginary values, the limit is -1.)
    (*b*) 0

2.3.2 (*a*) Has a nonremovable discontinuity at $z = 0$
    (*b*) Has a nonremovable discontinuity at $z = 1$
    (*c*) Continuous everywhere
    (*d*) Has a removable discontinuity at $z = 0$

2.3.4 The image of the disk under (*a*) is again the disk. The image of the disk under (*b*) is $\{w \in \mathbf{C}: |w - i| < |w - z|\}$. (Can you identify this set geometrically?)

2.3.5 Use the inequalities

$$|(u + iv) - (l_1 + il_2)| \leq |u - l_1| + |v - l_2|$$

$$|u - l_1| \leq |(u + iv) - (l_1 + il_2)|$$

and

$$|v - l_2| \leq |(u + iv) - (l_1 + il_2)|$$

2.4.2 Set $\zeta_n = z_n - z$. Then $\lim_{n \to 0} \zeta_n = 0$ and $Z_n = (z_1 + z_2 + \cdots + z_n)/n - z = (\zeta_1 + \zeta_2 + \cdots + \zeta_n)/n$. Let $\varepsilon > 0$ be given. Then there exists $n_1 \in \mathbf{N}$ such that $n > n_1$ implies $|\zeta_n| < \varepsilon/2$. Thus if $n \geq n_1$,

$$|Z_n| \leq \frac{|\zeta_1| + \cdots + |\zeta_{n_1 - 1}|}{n} + \frac{|\zeta_{n1}| + \cdots + |\zeta_n|}{n}$$

$$\leq \frac{|\zeta_1| + \cdots + |\zeta_{n_1 - 1}|}{n} + \frac{n - n_1}{n} \frac{\varepsilon}{2}$$

Choose $n_2 \geq n_1$ such that

$$\frac{|\zeta_1| + \cdots + |\zeta_{n_1 - 1}|}{n_2} < \frac{\varepsilon}{2}$$

Then for $n \geq n_2$

$$|Z_n| < \frac{\varepsilon}{2} + \left(1 - \frac{n_1}{n}\right) \frac{\varepsilon}{2} < \varepsilon$$

2.4.3 (*a*) 0; (*b*) does not exist; (*c*) 0; (*d*) 2; (*e*) $-\infty$.

2.4.4 As in calculus

2.4.5 As in calculus

2.4.6 As in calculus

2.4.7 (*a*) Absolutely convergent; (*b*) absolutely convergent; (*c*) absolutely convergent; (*d*) divergent; (*e*) conditionally convergent.

2.4.9 $\lim_{n \to \infty} \sup_{k \geq n} \sqrt[k]{|z_k|} = 1/\sqrt{2} < 1$

2.5.1 Let S be compact and let $\{z_n\}$ be a sequence in S. We must show that $\{z_n\}$ has a convergent subsequence. If $\{z_n\}$ has a finite number of distinct points, the assertion is true. If $\{z_n\}$ has infinitely many distinct points, use the Bolzano-Weierstrass theorem. Conversely, if every sequence in S has a convergent subsequence, we must show that S is compact. Show first that for each $\varepsilon > 0$ there exists a finite number of points $z_1, z_2, \ldots, z_n$ in S such that $S \subset$

$\bigcup_{k=1}^{n} N_\varepsilon(z_k)$. This shows that S is bounded. Use Exercise 2.2.6 to show that S is closed (or use the above fact and Lebesgue's covering lemma to show that every open covering of S has a finite subcovering).

2.5.2 From the definition of infimum, for each $n \in \mathbf{N}$ there exist $a_n \in A$ and $b_n \in B$ such that $d(A,B) \le |a_n - b_n| < d(A,B) + 1/n$. By Corollary 2.2, $\{a_n\}$ has a subsequence $\{a_{n_k}\}$ converging to some point $a \in A$. Show that $\{b_{n_k}\}$ has a subsequence converging to some point $b \in B$. Then clearly $d(A,B) = |a - b|$.

2.5.3 If f were not uniformly continuous, then there should exist $\varepsilon > 0$ such that for $n = 1, 2, \ldots$ and $\delta_n = 1/n$, there are points z_n and ζ_n such that $|z_n - \zeta_n| < 1/n$ and yet $|f(z_n) - f(\zeta_n)| \ge \varepsilon$. Apply Corollary 2.2 to the sequences $\{z_n\}$ and $\{\zeta_n\}$ to get a contradiction.

2.5.4 Show that $f[S]$ is closed and bounded.

2.6.3 (a) Open, bounded, arcwise connected, domain, star domain. Its boundary is the circle $|z - 3| = 1$. Its set of limit points is the closed disk $|z - 3| \le 1$, it has no isolated points, its closure is the closed disk $|z - 3| \le 1$, and its interior is the open disk $|z - 3| < 1$.

(d) Not open, not closed, bounded, arcwise connected, not a domain, not a star domain. Its boundary is $\{i\} \cup \{z \in \mathbf{C} : |z - i| = 1\}$, the set of all its limit points is the closed disk $|z - i| \le 1$, it has no isolated points, and its interior is the punctured disk $0 < |z - i| < 1$.

(j) Closed, unbounded, arcwise connected, and its boundary consists of the straight lines $x = \pm y$.

2.7.1 Observe that

$$\lim_{z \to \infty} \frac{a_n z^n + \cdots + a_0}{b_m z^m + \cdots + b_0} = \frac{a_n}{b_m} \lim_{z \to \infty} z^{n-m}$$

2.7.2 When S is considered as a subset of $\mathbf{C}$, it has no limit points, and so $S = S' \cup S$ is closed. But when S is considered as a subset of $\tilde{\mathbf{C}}$, it has ∞ as its limit point and $\infty \notin S$.

Chapter 3

3.2.1 (a) The domain of analyticity is $\mathbf{C} - \{0\}$ and the derivative is $3(z - 1/z)^2 \times (1 + 1/z^2)$.

3.2.2 By Example 3.4, a real-valued analytic function must be constant, and none of the four functions here is constant in any domain.

3.2.3 Proof is similar to Example 3.4.

3.2.5 Let $f(z) = z \operatorname{Re} z$. Then $[f(z) - f(0)]/z = (z \operatorname{Re} z)/z = \operatorname{Re} z \to 0$ as $z \to 0$. So f is differentiable at $z = 0$. Now if $z \ne 0$, show that

$$\lim_{h \to 0} \frac{f(z_0 + h) - f(z_0)}{h} = \begin{cases} z_0 + \operatorname{Re} z_0 & \text{if } h \text{ is real} \\ \operatorname{Re} z_0 & \text{if } h \text{ is pure-imaginary} \end{cases}$$

3.2.6 Let $P(z) = A(z - z_1) \cdots (z - z_n)$. Then

$$\frac{P'(z)}{P(z)} = \frac{1}{z - z_1} + \cdots + \frac{1}{z - z_n}$$

Let Re $z_k < 0$ for $k = 1, 2, \ldots, n$. Then for Re $z \geq 0$ we have

$$\text{Re} \, \frac{1}{z - z_k} = \text{Re} \, \frac{\bar{z} - \bar{z}_k}{|z - z_k|^2} > 0$$

Thus, Re $[P'(z)/P(z)] > 0$ for Re $z \geq 0$, which implies that $P'(z) \neq 0$ for Re $z \geq 0$. Hence the zeros of $P'(z)$ have negative real parts.

3.3.1 (a) 1; (b) ∞; (c) ∞; (d) ∞; (e) 1; (f) e.
3.3.2 (a) 1, converges at $z = -1$, diverges at $z = 1$
     (b) 1, diverges at every point on the circle of convergence
3.3.4 The series

$$\sum_{n=1}^{\infty} \frac{(0.9)^n}{n^4}$$

converges absolutely.
3.3.5 $1/z = 1/[1 + (z - 1)] = \sum_{n=0}^{\infty} (-1)^n (z - 1)^n$ for $|z - 1| < 1$. If $g(z) = 1/z$, then $1/z^2 = -g'(z) = \sum_{n=0}^{\infty} (-1)^{n+1}(n + 1)(z - 1)^n$ for $|z| < 1$ (where we obtain the power series for g' by differentiating term by term the power series for g).

3.4.1 (c) $e^{1 + (\pi/2)i} = e(\cos \pi/2 + i \sin \pi/2) = ie$
3.4.2 (a) If $f(z) = e^z$, then $f'(0) = \lim_{z \to 0} [(e^z - 1)/z]$.
     (b) If $g(z) = \sin z$, then $g'(0) = \lim_{z \to 0} [(\sin z)/z]$.
     (c) If $h(z) = \cos z$, then $h'(0) = \lim_{z \to 0} [(\cos z - 1)/z]$.
3.4.3 (a) If $z = x + iy$, then $-1 + i = e^{x+iy} = e^x(\cos y + i \sin y)$. Thus $e^x \cos y = -1$ and $e^x \sin y = 1$. By squaring both sides of these equations and adding, we get $e^{2x} = 2$. Hence $2x = \ln 2$ and $x = \frac{1}{2} \ln 2 = \ln \sqrt{2}$. Also (since $e^x = \sqrt{2}$) we know that $\sqrt{2} \cos y = -1$ and $\sqrt{2} \sin y = 1$. Thus, $y = 3\pi/4 + 2\pi n$ for $n = 0, \pm 1, \pm 2, \ldots$. So $z = \ln \sqrt{2} + i(3\pi/4 + 2\pi n)$ for $n = 0, \pm 1, \pm 2, \ldots$.
     (b) $z = i(\pi/2 + 2n\pi)$ for $n = 0, \pm 1, \pm 2, \ldots$
     (c) It has no root.
3.4.4 (a) Use Property 20 of the text. (b) Use Exercise 3.4.4a.
3.4.8 Use the method of Example 3.7 [or observe that $\cos z = \sin (\pi/2 - z)$].

3.4.12 (a) $|e^z - 1| = \left| \dfrac{z}{1!} + \dfrac{z^2}{2!} + \cdots \right| \leq |z| \left(1 + \dfrac{1}{2!} + \dfrac{1}{3!} \right) + \cdots$

$$= |z|(e - 1) < \tfrac{7}{4}|z|.$$

Also,

$$|e^z - 1| > |z| \left(1 - \frac{1}{2!} - \frac{1}{3!} - \cdots \right) = |z|(3 - e) > \frac{1}{4}|z|$$

3.5.1 (a) Log $\sqrt{2} + i(\pi/4 + 2\pi k)$ for $k = 0, \pm 1, \pm 2, \ldots$
     (b) Log $(1 + i)^2 = $ Log $2i = $ Log $2 + i(\pi/2)$
     (c) $\log e = 1 + 2\pi i k$ for $k = 0, \pm 1, \pm 2, \ldots$
     (d) Log $e = 1$

3.6.4 Use the Cauchy-Riemann equations.
3.6.5 Use Exercise 3.6.4a.

3.6.8 Show first that

$$u_x = \frac{\partial u}{\partial r} \cos \theta - \frac{1}{r} \frac{\partial u}{\partial \theta} \sin \theta$$

$$u_y = \frac{\partial u}{\partial r} \sin \theta + \frac{1}{r} \frac{\partial u}{\partial \theta} \cos \theta$$

$$v_x = \frac{\partial v}{\partial r} \cos \theta - \frac{1}{r} \frac{\partial v}{\partial \theta} \sin \theta$$

$$v_y = \frac{\partial v}{\partial r} \sin \theta + \frac{1}{r} \frac{\partial v}{\partial \theta} \cos \theta$$

3.7.1 (a) $3x^2 y - y^3 + y + c$; (b) $-e^x \cos y + c$

3.7.2 $a = -1$, $v(x, y) = 2xy + 3$

3.7.4 Show that v is harmonic and u and v satisfy the Cauchy-Riemann equations.

3.7.5 Show that $(u_y - v_x)_x = (u_x + v_y)_y$ and $(u_y - v_x)_y = -(u_x + v_y)_x$. (Remember that $u_{xx} + u_{yy} = 0$ and $v_{xx} + v_{yy} = 0$.)

Chapter 4

4.2.1 $(\gamma_1 + \tilde{\gamma}_2)(t) = \begin{cases} z_0 e^{it} & 0 \le t \le \pi/4 \\ z_0 e^{-i(2\pi - t)} & \pi/4 \le t \le 2\pi \end{cases}$

4.2.2 (a) $\displaystyle\int_{\gamma_1} x \, dz = \int_0^1 t(1 + i) \, dt = (1 + i)/2$

(b) $\displaystyle\int_{\gamma_2} x \, dz = \int_0^{\pi/2} (1 - \cos t)(\sin t + i \cos t) \, dt$

$$= -\cos t + i \sin t + \tfrac{1}{2} \cos^2 t - i[t/2 - (\sin 2t)/4]\big|_0^{\pi/2}$$
$$= \tfrac{1}{2} + i(1 - \pi/4)$$

(c) $\displaystyle\int_{\gamma_3} x \, dz = \int_0^1 t \, dt + \int_1^2 i \, dt = \tfrac{1}{2} + i$

4.2.3 Use Example 4.1.

4.2.4 Use Example 4.1.

4.2.5 $\displaystyle\left| \int_\gamma \frac{dz}{4 + 3z} \right| = \left| \int_0^{2\pi} \frac{ie^{it} \, dt}{4 + 3e^{it}} \right| \le \int_0^{2\pi} \frac{dt}{|(4 + 3 \cos t) + 3i \sin t|}$

$$= \int_0^{2\pi} \frac{dt}{\sqrt{25 + 24 \cos t}}$$

$$\le \int_0^{\pi/2} \cdots + \int_{\pi/2}^{3\pi/2} \cdots + \int_{3\pi/2}^2 \cdots$$

$$\le \frac{1}{5} \frac{\pi}{2} + 1\pi + \frac{1}{5} \frac{\pi}{2} = \frac{6\pi}{5}$$

4.3.2 Let γ be the contour shown in the figure.

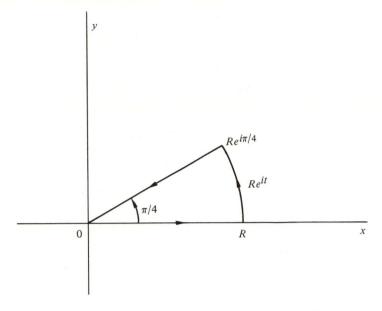

By applying Theorem 4.3 to the entire function $f(z) = e^{iz^2}$, we have

$$0 = \int_\gamma e^{iz^2}\, dz = \int_0^R e^{ix^2}\, dx + \int_0^{\pi/4} e^{iR^2e^{2it}} iRe^{it}\, dt + \int_R^0 e^{it^2 e^{i\pi/2}} e^{i\pi/4}\, dt$$

$$= J_1 + J_2 + J_3$$

Use the inequality $\sin t \geq 2t/\pi$ for $0 \leq t \leq \pi/2$ to show that $J_2 \to 0$ as $R \to \infty$.

4.4.1 Find constants A, B, C such that

$$\frac{1}{z(z^2 + 1)} = \frac{A}{z} + \frac{B}{z + i} + \frac{C}{z - i}$$

and then use formula (4.12).

4.4.2 Decompose $1/[z^2(z - 1)]$ into the sum of partial fractions.

4.4.3 (a) $-12\pi i$; (b) $-2\pi i/6!$; (c) decompose $1/(z^4 - 1)$ into the sum of partial fractions.

4.4.4 The entire function $e^{if(z)}$ is bounded by e^M and so is constant. Thus f is constant.

4.4.6 Use formula (4.18).

4.4.7 $f(z_0) = \dfrac{1}{2\pi i} \displaystyle\int_\gamma \frac{f(\zeta)}{\zeta - z_0}\, d\zeta = \dfrac{1}{2\pi i} \displaystyle\int_0^{2\pi} \frac{f(z_0 + re^{i\theta})}{re^{i\theta}} ire^{i\theta}\, d\theta$

4.4.8 Evaluate $\int_{|z|=1} e^z/z\, dz$ first by using Theorem 4.4 and then by setting $z = e^{i\theta}$ for $0 \leq \theta \leq 2\pi$.

4.5.1 (a) $\displaystyle\sum_{n=0}^{\infty} (-1)^n \frac{z^{2n+1}}{(2n+1)!}$ for all $z \in \mathbf{C}$

(b) $\displaystyle\sum_{n=1}^{\infty} (-1)^{n+1} \frac{(z-\pi)^{2n}}{2n!}$ for all $z \in \mathbf{C}$

(d) $\displaystyle\sum_{n=0}^{\infty} \frac{(-1)^n}{i^{n+1}} (z-i)^n$ for $|z - i| < 1$

(f) $(-6 + 8i) - (12 + 4i)(z + 2i) - 6i(z + 2i)^2 + (z + 2i)^3$
$\qquad\qquad\qquad\qquad\qquad\qquad\qquad\qquad\qquad$ for all $z \in \mathbf{C}$

4.5.2 $|f|$ clearly attains a maximum on $\bar{D}$. Since by Theorem 4.13 $|f|$ does not attain a maximum on D, it follows that it attains a maximum on the boundary of D.

4.5.3 $M > |f(z)| = |z^2 g(z)| = |z|^2 |g(z)| = |g(z)|$ on $|z| = 1$. Thus $|f(z)| = |z^2 g(z)| = |z|^2 |g(z)| = \frac{1}{4}|g(z)| < \frac{1}{4}M$ on $|z| = \frac{1}{2}$. Since by Exercise 4.5.2 the maximum of $|f|$ on $|z| \le \frac{1}{2}$ is attained on $|z| = \frac{1}{2}$, the result follows.

4.5.4 Apply Theorem 4.13 and Corollary 4.5a to the function
$$f(z) = (z - z_1)(z - z_2) \cdots (z - z_n) \qquad \text{on } D$$

4.5.5 Use Corollary 4.3 with $z_0 = 0$.

4.5.6 Find the roots of $z^n = 1$ and factor $z^n - 1$ to get
$$z^n - 1 = (z - 1)(z - e^{2\pi i/n})(z - e^{4\pi i/n}) \cdots (z - e^{2(n-1)\pi i/n})$$

Thus for $z \ne 1$ we have
$$(z - e^{2\pi i/n})(z - e^{4\pi i/n}) \cdots (z - e^{2(n-1)\pi i/n})$$
$$= \frac{z^n - 1}{z - 1} = z^{n-1} + z^{n-2} + \cdots + z + 1$$

Taking limits as $z \to 1$ we get
$$(1 - e^{2\pi i/n})(1 - e^{4\pi i/n}) \cdots (1 - e^{2(n-1)\pi i/n}) = n$$

Multiply this identity by its conjugate and use the fact that $\cos 2a = 1 - 2\sin^2 a$.

4.6.1 It is sufficient to note that the exercise can be reduced to the case where f and g are power series on D and then prove it for this case.

4.6.2 Let $\gamma(t) = 2\cos t + 3i\sin t$ for $0 \le t \le 2\pi$. Then γ is the required curve. As γ is homotopic through closed curves in $\mathbf{C} - \{0\}$ to the curve $|z| = 1$, we see that $\int_\gamma dz/z = \int_{|z|=1} dz/z = 2\pi i$. The conclusion follows by integrating $\int_\gamma dz/z$ using the technique developed in Sec. 4.2.

4.7.1 $I(\gamma, \frac{1}{2}) = 1$, $I(\gamma, 2 + i/2) = 0$, $I(\gamma, -3 + i) = 0$

4.7.2 Use Exercise 4.7.1 and Corollary 4.16.

Chapter 5

5.2.1 (a) $\displaystyle\sum_{n=0}^{\infty} \frac{(-1)^n}{2^n}(z - 1)^n + (-1 + 2i)\sum_{n=0}^{\infty} (-1)^n i^n (z - 1)^{-n-1}$
$$- (1 + 2i)\sum_{n=0}^{\infty} i^n (z - 1)^{-n-1} \qquad \text{if } 1 < |z| < 2$$

(b) $\displaystyle\sum_{n=0}^{\infty} (-1)^n(2^n - 3^n)(z - 1)^{-n-1}$ if $|z - 1| > 3$

(c) $\displaystyle -\frac{1}{2(z - 1)^2} - \frac{1}{4(z - 1)} - \frac{1}{3}\sum_{n=0}^{\infty} (z - 1)^n - \sum_{n=0}^{\infty} \frac{(-1)^n}{2^n}(z - 1)^n$

$\qquad\qquad\qquad\qquad\qquad\qquad\qquad\qquad$ if $0 < |z - 1| < 1$

5.2.2 (a) $\displaystyle\frac{1}{z^7} + \frac{1}{3!}\frac{1}{z^5} + \frac{1}{5!}\frac{1}{z^3} + \frac{1}{7!}\frac{1}{z} + \frac{1}{9!}z + \cdots + \frac{1}{(2n + 9)!}z^{2n+1} + \cdots$

(b) $\displaystyle\sum_{n=0}^{\infty} (z^{2n-3}/n!)$

5.3.1 (a) $z = n\pi$, for $n = 0, \pm 1, \pm 2, \ldots$, are simple poles.
(b) $z = 0$ is a pole of order 4.
(c) $z = 2\pi i n$, for $n = 0, \pm 1, \pm 2, \ldots$, are simple poles.
(d) $z = 0$ is an essential singularity.

5.3.3 Define $f(0) = {}^1\!/_2$.

5.4.1 $4\pi i$

5.4.2 0

5.5.1 (a) $(2\pi/\sqrt{3})(2 - \sqrt{3})^3$
(b) $3\pi\sqrt{2}$
(c) $\pi/(2e^3)$

5.5.3 (a) $-\pi/(8e^\pi)$
(b) $\pi/(2e)$
(c) $[\pi/(6e^2)](2e - 1)$

5.5.5 (a) $\pi\sqrt{2}/4$
(b) $\pi\sqrt{3}/6$
(c) $2\pi\sqrt{3}/3$

5.6.1 (a) If $f(z) = z^3 - 2$, then

$$\int_{|z|=3} \frac{z^2}{z^3 - 2}\, dz = \frac{1}{3}\int_{|z|=3} \frac{f'(z)}{f(z)}\, dz = \frac{1}{3}\cdot 2\pi i \cdot 3 = 2\pi i$$

(All the zeros of f lie inside $|z| = 3$.)
(b) If $f(z) = z^4 + 2z + 1$, then

$$\int_{|z|=2} \frac{2z^3 + 1}{z^4 + 2z + 1}\, dz = \frac{1}{2}\int_{|z|=2} \frac{f'(z)}{f(z)}\, dz = \frac{1}{2}\cdot 2\pi i \cdot 4 = 4\pi i$$

[Since for $|z| \geq 2$, $|z^4 + 2z + 1| \geq |z^4 + 2z| - 1 = |z|(|z|^3 - 2) - 1 \geq 2(8 - 2) - 1 = 11 > 0$, all the zeros of f lie inside $|z| = 2$.]

5.6.2 Let $g(z) = e^z$, $f(z) = -3z^7$ and apply Rouché's theorem.

5.6.3 One. [Take $f(z) = z$ and $g(z) = e^{z-2}$ and apply Rouché's theorem.]

5.6.4 Let $P(z) = a_n z^n + a_{n-1}z^{n-1} + \cdots + a_1 z + a_0$. Take $f(z) = a_n z^n$ and $g(z) = a_{n-1}z^{n-1} + \cdots + a_1 z + a_0$.

INDEX

Photo Credits